Certified in Emerging Technology

CLOUD
FUNDAMENTALS
Study Guide

Part of the CET Stackable Certificate Program
www.isaca.org/cet

ISACA®

About ISACA

For more than 50 years, ISACA® (*www.isaca.org*) has advanced the best talent, expertise and learning in technology. ISACA equips individuals with knowledge, credentials, education and community to progress their careers and transform their organizations, and enables enterprises to train and build quality teams. Among those credentials, ISACA advances and validates business-critical skills and knowledge through the globally respected Certified Information Systems Auditor® (CISA®), Certified in Risk and Information Systems Control® (CRISC®), Certified Information Security Manager® (CISM®), Certified in the Governance of Enterprise IT® (CGEIT®) and Certified Data Privacy Solutions Engineer™ (CDPSE™) credentials. ISACA is a global professional association and learning organization that leverages the expertise of its 145,000 members who work in information security, governance, assurance, risk and privacy to drive innovation through technology. It has a presence in 188 countries, including more than 220 chapters worldwide.

Disclaimer

Reservation of Rights

ISACA
1700 E. Golf Road, Suite 400
Schaumburg, IL 60173, USA
Phone: +1.847.660.5505
Fax: +1.847.253.1755
Contact us: https://support.isaca.org
Website: www.isaca.org

Provide Feedback: To comment, please visit https://support.isaca.org.
Participate in the ISACA Online Forums: https://engage.isaca.org/onlineforums

Twitter: http://twitter.com/ISACANews
LinkedIn: www.linkedin.com/company/isaca
Facebook: www.facebook.com/ISACAGlobal
Instagram: www.instagram.com/isacanews/

ISBN 978-1-60420-838-2
Cloud Fundamentals Study Guide
Printed in the United States of America

Acknowledgments

The *Cloud Fundamentals Study Guide* is the result of the collective efforts of many volunteers. Special thanks go to:

Expert Reviewers

Marianne Azer, Ph.D., Member of the Egyptian Parliament, Egypt
Vladlena Benson, Ph.D., CDPSE, United Kingdom
Earnest Collins, CISA, CISM, CFE, CISSP, AES, USA
Ramsés Gallego, CISM, CGEIT, CCSK, CISSP, Six Sigma Black Belt, Micro Focus, Spain
Niel Harper, CISA, CRISC, CDPSE, CISSP, Denmark
Luis Jugo, CISA, CCSA, CIA, CRMA, PMP, USA
Neil Lappage, CISM, CISSP, Australia
Tuan Phan, CISSP, CBSP, Security+, PMP, Caplock Security, USA
Ron Quaranta, EC–Information Systems and Technology Project Management, Wall Street Blockchain Alliance, USA

Board of Directors

Tracey Dedrick, Chair, Former Chief Risk Officer, Hudson City Bancorp, USA
Rolf von Roessing, Vice-Chair, CISA, CISM, CGEIT, CDPSE, CISSP, FBCI, Partner, FORFA Consulting AG, Switzerland
Gabriela Hernandez-Cardoso, Independent Board Member, Mexico
Pam Nigro, CISA, CRISC, CGEIT, CRMA, Vice President–Information Technology, Security Officer, Home Access Health, USA
Maureen O'Connell, Board Chair, Acacia Research (NASDAQ), Former Chief Financial Officer and Chief Administration Officer, Scholastic, Inc., USA
David Samuelson, Chief Executive Officer, ISACA, USA
Gerrard Schmid, President and Chief Executive Officer, Diebold Nixdorf, USA
Gregory Touhill, CISM, CISSP, President, AppGate Federal Group, USA
Asaf Weisberg, CISA, CRISC, CISM, CGEIT, Chief Executive Officer, introSight Ltd., Israel
Anna Yip, Chief Executive Officer, SmarTone Telecommunications Limited, Hong Kong
Brennan P. Baybeck, CISA, CRISC, CISM, CISSP, ISACA Board Chair, 2019-2020, Vice President and Chief Information Security Officer for Customer Services, Oracle Corporation, USA
Rob Clyde, CISM, ISACA Board Chair, 2018-2019, Independent Director, Titus, and Executive Chair, White Cloud Security, USA
Chris K. Dimitriadis, Ph.D., CISA, CRISC, CISM, ISACA Board Chair, 2015-2017, Group Chief Executive Officer, INTRALOT, Greece

Page intentionally left blank

Table of Contents

LIST OF FIGURES

Chapter 1. Cloud Computing Concepts

Chapter 2. Cloud Service Models

Chapter 3. Cloud Governance

Chapter 4. Cloud Service Supports

Page intentionally left blank

An Introduction to Cloud Computing

Few would argue that, over the past few decades, technology advances have been introduced at record speed and adopted with equal alacrity. Indeed, a failure to adopt new technologies can be the harbinger of decline for enterprises, as their competitors leverage technology to outcreate, outdesign, outproduce and outmarket them.

A technology that has already overhauled many enterprises' infrastructure and processing is cloud computing. While it is not necessarily the solution for every enterprise's challenges—some enterprises prefer to maintain their data and processing on-premise—many organizations have recognized the benefits it can provide and are quickly adopting its offerings. Enterprises—and individuals who seek to play a major role in their enterprise's success—would do well to understand what cloud computing is, how it works, how it can benefit an organization and what risk it contains.

The *Cloud Fundamentals Study Guide* works through each aspect of cloud computing, its characteristics, common decision points, gaps and security vulnerabilities. It helps individuals prepare for the ISACA® Fundamentals certificate exam, one of the components of the ISACA Certified in Emerging Technology™ certification program.

Every chapter contains the following:

- **Learning objectives**—Knowledge that learners should understand after reading the chapter
- **Sections**—Knowledge content
- **Knowledge checks**—Quiz on the chapter knowledge

Page intentionally left blank

Chapter 1:

Cloud Computing Concepts

Cloud Computing Concepts

Cloud Computing Concepts

1.1 Learning Objectives

After completing this chapter, learners will be able to:

1. Define cloud technology.
2. Articulate the cloud service model characteristics and capabilities.
3. Identify the components that make up cloud technology.
4. Explain what considerations need to be made for cloud service implementation.
5. Distinguish different cloud deployment models.

1.2 Overview

In many countries, such as Australia, government agencies have an explicit obligation to consider cloud services when procuring new information and communication technology (ICT) requirements for their test and development needs, and to migrate public-facing websites to public cloud services. The agencies must choose cloud services when they represent the best value and adequate risk management compared to other available options.

As defined by the US National Institute of Standards and Technology (NIST), cloud computing is a "model for enabling convenient, on-demand network access to a shared pool of configurable computing resources, such as, networks, servers, storage, applications, and services, that can be rapidly provisioned and released with minimal management effort or service provider interaction."[1]

Adding computer virtualization to the NIST definition (on-demand computer resources requiring minimal management effort) creates a cloud computing model that offers enterprises virtual processing power in a variety of possible implementations.

While the models and types of cloud services vary significantly, they share five essential characteristics:

- **On-demand self-service**—Computing capabilities can be provisioned without human interaction from the service provider.
- **Broad network access**—Computing capabilities are available over the network and can be accessed by diverse client platforms.
- **Resource pooling**—Computer resources are pooled to support a multi-tenant model.
- **Rapid elasticity**—Resources can scale up or down rapidly and, in some cases, automatically, in response to business demands.
- **Measured service**—Resource utilization can be optimized by leveraging charge-per-use capabilities.

While cloud computing is often categorized as exclusively an IT issue, such thinking underplays its potential impact. It also represents a weighty business decision for an organization, one that requires consideration and input from many stakeholders. These stakeholders and the enterprises they support are often drawn to the potential benefits of cloud computing, summarized in **figure 1.1**.

[1] Mell, P.; T. Grance; Special Publication (SP) 800-145 (Draft), "The NIST Definition of Cloud Computing," National Institute of Standards and Technology (NIST), USA, 2011, p. 3, *faculty.winthrop.edu/domanm/csci411/Handouts/NIST.pdf*

Figure 1.1—Cloud Benefits	
Benefit	**Description**
Tangible	
Cost reduction	Computing cost is shifted from capital expenditure to operational cost because the cloud provider supplies the underlying infrastructure as part of the service bundle. In addition, the cloud promises cost reduction in the following areas: • Labor (IT system administration hours/headcount) • Application software (Software as a Service [SaaS] only) • Licensing purchase and maintenance • Technical support and user support • Maintenance (upgrades, updates, patches, etc.) • Hosting (physical building, power, cooling, etc.)
Enhanced productivity	User mobility and ubiquitous access can increase productivity. Collaborative applications increase productivity and reduce rework.
Optimized resource utilization	Enterprises use only the computing resources they need, thus reducing system idle time waste.
Improved security/ compliance	Cloud providers may offer robust security controls as a market differentiation.
Access to skills and capabilities	Customers benefit from top-notch skills and capabilities while avoiding employment costs (recruiting, salary, benefits, training, etc.)
Scalability	On-demand provisions or computing resources eliminate the cost of capacity planning.
Agility	Agility contributes to cost reduction and productivity enhancement due to faster provisioning of systems: • Faster application deployment (SaaS) • Faster application development/testing (Platform as a Service [PaaS])
Customer satisfaction	Effective utilization of cloud applications can increase collaboration between the enterprise and its customers or reduce response time to customer inquiries.
Reliability	Cloud providers have redundant sites that can address business continuity and disaster recovery in a more efficient manner.
Performance	Better performance and uptime can result from continuous and consistent operations monitoring by the cloud provider.
Intangible	
New business opportunities	A cloud application (SaaS) may be the critical element to land a new business or expand into new markets.
Focus on core business	IT resources can be allocated to support core business functions.
Employee satisfaction/ innovation	Mobility and faster performance can improve employee satisfaction and boost innovation.
Collaboration	Real-time collaboration can increase quality and innovation.
Risk transfer	Some risk (e.g., security breaches, data loss, disaster recovery) can be transferred to the cloud service provider (CSP); this could represent a tangible or intangible benefit.

Source: ISACA, *Controls and Assurance in the Cloud: Using COBIT® 5*, USA, 2014, *https://www.isaca.org/bookstore/cobit-5/cb5ca*

Depending on business needs, any or all of these benefits could be a sufficient reason to consider a cloud computing solution. However, cloud services can also generate significant risk, and any enterprise contemplating cloud usage must be prepared to address it. If not handled well, the risk can quickly turn the cloud experience into an information security management nightmare derived from the loss of controls over physical and logical assets.

1.3 Components of the Cloud

Virtualization. Hypervisors. Containers. These are just some of the terms encountered when entering the cloud computing space. Because there is overlap among them, it is easy to assume they are all basically the same thing, performing the same function. However, that is an oversimplification.

The ISACA Glossary defines virtualization as "[t]he process of adding a 'guest application' and data onto a 'virtual server,' recognizing that the guest application will ultimately part company from this physical server." Other definitions cite its use of software on a single, physical hardware system to create a virtual, rather than actual, version of something, such as an operating system, storage, a server or a network. "Virtual" is the opposite of "physical," not of "real." It is a technology, not an environment.

In enterprise information technology (IT), virtualization alters the technical architecture because it allows different resources to be executed in a single (or multilayer) environment. In general, it turns one piece of hardware into the host for many other pieces and, consequently, over time, reduces the enterprise's capital expense, administration cost and other financial costs on the profit and loss statement.

Virtualization, as a term and a concept, has broad utility and can be applied to several areas: virtualized servers, virtualized storage, virtualized processors, virtual memory, virtual desktops, virtualized network, etc. Because of its extensive applications and cost savings, it is being evaluated by chief information officers (CIOs) worldwide as they strategize how to provide agility and computing power to meet their enterprise needs. In addition, because organizations today require a quick and reliable way to provision technical resources that enable a faster time to market, virtualization is on the C-level agenda and is already enhancing the effectiveness of many enterprises around the globe.

Originally virtualization was used primarily to facilitate server consolidation, but now many other approaches present themselves. Virtualization starts during the technical environment design phase, when the design team considers how to support the business processes and identifies assets needed to convert the plan to reality. During this phase, enterprises often realize that the life cycle of provisioning the right hardware and other equipment can go on longer than expected and there is a faster way to deploy them: by building in some abstraction from the physical world and hosting different virtual resources within the boundaries of a unique physical resource.

There are many benefits of implementing virtualization within enterprise IT, as itemized below and summarized in **figure 1.2**.

- **Cost reduction**—By consolidating many instances of (virtualized) servers on a physical one, enterprises lower their hardware expenditures. In addition to lower capital expenditures, virtualized environments enable enterprises to save on maintenance and energy, often resulting in a reduced total cost of ownership.
- **Automation**—Technology allows some virtualized environments to be provisioned as needed and on the fly, thus facilitating automation of business processes and eliminating the need to continually resource and manage portions of the technical environment that support sporadic business needs. Some virtualization technology facilitates the automatic allocation of a process for its optimal performance within a pool of virtualized environments.
- **Responsiveness**—Since the virtual environment can provision itself to get the best out of available resources, response times are faster and downtimes can be reduced to near zero, improving agility and performance.
- **Decoupling**—Processes that once needed to exist within the same physical machine can now be easily separated while still maintaining the robustness and security required. The different virtualized worlds (network, operating system [OS], database, application, etc.) can be decoupled (even distributed in different geographies) without threatening integrity in the process.
- **Flexibility**—The relatively easy creation or preparation of the right environment for the right application enables enterprises to provide flexibility to the infrastructure, not only in the test or preproduction phases but also in the

production area. When a new procedure or technical/business requirement arises, virtualization's ability to enable rapid creation of the environment allows the business to test the environment without having to wait for the regular provisioning process to be executed and delivered.

- **Agility**—Agility facilitates quick adaptation to business needs—for example, if orders peak and additional computing power is needed. An enterprise may even choose to overcommit the resources of a physical machine, since virtualization facilitates rapid movement of the different resources that "live" in one physical machine to other virtual machines. In this way, virtualization supports alignment with business needs.
- **Workload balancing**—Deploying several virtual environments guarantees the good practices of high availability, redundancy and failover since workloads can go where they are more efficient. Thus, virtualization focuses not only on effectiveness (doing the right things) but also efficiency (doing things in a faster, cheaper and more reliable way).
- **Simplification**—Virtual IT is still IT, so some of the typical IT difficulties exist even within a virtual environment. However, reducing the number of physical servers significantly reduces the probability of failure and the cost of management and results in simplification—one of the promises of virtualization.
- **Space utilization**—Server consolidation saves space in the data center and facilitates scalability since many servers exist within one server.
- **Sustainability**—Virtualized environments use less environmental resources. Energy consumption in data centers is often wasted on machines that are consistently underutilized. Since virtualization allows for many virtual machines to run on one physical machine, less energy is needed to power and cool devices.

Figure 1.2—Five Reasons to Virtualize

Outcome	How It Is Achieved
1. Reduce IT complexity.	Applications and their operations systems are encapsulated in virtual machines that are defined in software, making them easy to provision and manage.
2. Enable standardization.	Since applications are decoupled from hardware, the data center may converge on a narrower range of hardware devices.
3. Improve agility.	Applications and virtual machines can be copied and moved in real time—and in the cloud—in response to changing business conditions.
4. Improve cost efficiency.	Virtual machines can easily be moved to consume spare capacity wherever it exists, thus generating more work from less hardware.
5. Facilitate automation.	Virtual infrastructure is easily provisioned and orchestrated by software-driven processes, especially when the underlying hardware is standardized.

As with any technology, there is risk associated with virtualization. The risk can be categorized into three groups:

- **Attacks on virtualization infrastructure**—There are two primary types of attacks on virtualization infrastructure: hyperjacking and virtual machine (VM) jumping (or guest-hopping).
- **Attacks on virtualization features**—Although there are multiple features of virtualization that can be targeted for exploitation, the more common targets include VM migration and virtual networking functions.
- **Compliance and management challenges**—Compliance auditing and enforcement, as well as day-to-day system management, are challenging issues when dealing with virtualized systems. VM sprawl introduces a challenge to the enterprise: Because VMs are much easier to provision and deploy than physical systems, the number and types of VMs can easily get out of hand. Furthermore, VM provisioning is often administered by different groups within an organization, making it difficult for the IT function to control what applications, OSs and data are deployed to them.

A hypervisor is computer software that allows various software applications running on different operating systems to coexist on the same server at the same time. The hypervisor and the guests it supports are software and, as such, need to be regularly patched and hardened. Cloud infrastructures are based on hypervisors and are controlled through a central hypervisor manager or client. They provide the link between virtual machines and the underlying physical resources required to run the machines by using hypercalls (similar to system calls, but for virtualized systems). As such, they are vital for cloud virtualization.

A container is a "virtual runtime environment that runs on top of a single operating system (OS) kernel and emulates an operating system rather than the underlying hardware."[2] Virtualization provides an abstract machine; a container provides an abstract operating system. Therefore, applications that are running in a container share an operating system, while VM systems can run different operating systems.

In summary, these technologies constitute a kind of "evolution" of the cloud. Cloud computing (as it is known today) sprung from these technologies and their availability. Virtualization of hardware paved the way for a single system being used for multiple tasks and purposes. Hypervisors are the software that can be used to virtualize depending on the application. Containers take it a step further and allow for minimal resource use to run full services. Together, they are used with automation, scaling and redundancy to provide the technical cogs of the cloud.

1.4 Cloud Service Implementation Considerations

In preparing to implement cloud solutions, an enterprise must consider some of the challenges that may arise. For example, the cloud computing implementation may conflict with the enterprise's culture. The cloud solution may struggle to accommodate changes in requirements identified after implementation. The implementation may have failed to recognize and establish controls to offset known and future cloud risk. Enterprise and stakeholder expectations for the implementation may not have been thoroughly understood or communicated.

An effective implementation may require the establishment of an implementation team consisting of appropriate members, creating a trusted environment and establishing common goals and effectiveness measures. The team members—and anyone with a role to play in implementation—must be empowered to do their best work by ensuring that accountabilities are assigned, training is provided, and organizational structures and human resource (HR) processes are aligned. The team should be encouraged to identify and communicate short-term wins that maintain motivation and support change enablement. While the team members cannot be expected to have the answers and solutions to all issues that arise before and during implementation, among the many questions they must consider are:

- Which security policies apply within the enterprise? Which regulatory restrictions apply to the enterprise and to any locations where a CSP might reside?
- How will moving to the cloud influence the enterprise's processes? Which processes depend on assets that could move to the cloud? Are these processes considered to be critical for the business?
- How will the relationship with the CSP be managed? How are roles and responsibilities defined?
- How will change within the enterprise be managed? How can an information culture be imposed upon the CSP?
- Which assets are considered for cloud computing?
- Which service capabilities are expected of the CSP? How will performance be measured? How will issues be reported?
- Which skills and competencies are required to manage the enterprise assets? Does the enterprise wish to keep these in-house after a decision has been made to move to the cloud?

Because the amount of work and the complexity involved in preparing for and carrying out a successful implementation, some enterprises may choose to engage professional services to assist the in-house team in managing the transition to the cloud.

[2] Firesmith, D.; "Virtualization via Containers," Carnegie Mellon University, Software Engineering Institute, 25 September 2017, *insights.sei.cmu.edu/sei_blog/2017/09/virtualization-via-containers.html#:~:text=Note%20that%20virtualization%20via%20containers,rather%20than%20the%20underlying%20hardware*

Migrating the data is a significant aspect of cloud implementation. Consideration of migration issues should be addressed as early as the drafting of the service contract terms. The contract should provide for continuity of accessibility, usability and preservation of all enterprise data, regardless of any migration of data to other formats during the contract. Terms should provide for appropriate testing to ensure data integrity prior to any migration. It should also include provisions relating to the migration of data to new formats, when appropriate, and the provision of proper documentation about migration activities to the enterprise. Because even the best of business relationships sometimes come to an end, the contract should also define provisions relating to the migration of data on termination of the contract.

Successful migration begins with a plan that aligns with the enterprise's program and project portfolios. Close coordination among all facets of migration will help ensure that value is delivered and the required resources are available when needed to perform the necessary work. Preparation for migration should be an integral part of the enterprise's development methods and may include audit trails and a recovery plan should the migration fail.

The impact of migration depends on the cloud service model and deployment model being used, as they will indicate an appropriate balance for protection of organizational assets during migration. The enterprise should classify its assets into categories for an optimal selection of cloud arrangements. Generally, data can be classified as public, restricted, for internal use, secret and top-secret. A data life-cycle process can also be defined.

Two good rules of thumb, regardless of service model and deployment selected, are:

- Prior to migration, encrypt all sensitive assets being migrated to the CSP to prevent disclosure.
- Test changes independently in accordance with a defined test plan prior to migration to the live operational environment.

A post-implementation review should be conducted to confirm the outcome and results of the cloud implementation, identify lessons learned and develop action plans for future implementations. The review should also evaluate and check the actual performance and outcomes of the cloud solution against the predicted performance and outcomes and identify opportunities for improvement.

1.5 Cloud Deployment Models

Cloud customers have their choice of four cloud deployment models (see **figure 1.3**):

- **Private cloud**—One client (one individual or one enterprise). Several departments or divisions may be represented, but all exist within the same enterprise. Private clouds often employ virtualization within an enterprise's existing computer servers to improve computer utilization. A private cloud also typically involves provisioning and metering components, enabling rapid deployment and chargeback where appropriate. This model is most closely related to traditional IT outsourcing models in the marketplace, but it can also be an enterprise's internal delivery model. Because of this, the private model tends to offer the least amount of new risk and security challenge.
- **Public cloud**—An offering from one CSP to many clients who share the cloud processing power and storage capacity concurrently. Public cloud clients share applications, processing power and data storage space communally. Client data are commingled, but segregated, for example, through the use of metatags. This model tends to present the greatest assurance challenges for security and risk managers.
- **Community cloud**—A private-public cloud with clients who have a common connection or affiliation, such as a trade association, industry or locality. The community cloud business model allows a CSP to provide cloud tools and applications that are specific to the needs of the community.
- **Hybrid cloud**—A combination of two or more of the previously defined deployment models. Each of the three cloud deployment models has specific advantages and disadvantages that are relative to the other deployment

models. A hybrid cloud leverages the advantage of the other cloud models, providing a more optimal user experience.

Figure 1.3—Cloud Deployment

Deployment	Description	Examples
Private cloud	• Operated solely for one enterprise • May be managed by the enterprise or a third party • May exist on- or off-premise	• Hewlett Packard enterprise (HPE) • VMWare • Oracle
Public cloud	• Made available to the general public or a large industry group • Owned by an organization selling cloud services	• Amazon Web Services (AWS) • Microsoft (Azure) • Google Cloud Platform
Community cloud	• Shared by several enterprises • Supports a specific community that has a shared mission or interest • May be managed by the enterprises or a third party	• eCommerce • HR • Manufacturing
Hybrid cloud	A combination of two or more cloud deployment models (private, community or public) that remain unique entities, but are bound together by standardized or proprietary technology that enables data and application portability, e.g., cloud bursting for load balancing between clouds	

Source: ISACA, *Controls and Assurance in the Cloud: Using COBIT® 5*, USA, 2014, *https://www.isaca.org/bookstore/cobit-5/cb5ca*

The costs of services vary across the different deployment models. Private cloud deployments are often the costliest option; public clouds, the least. The hybrid may offer a reasonable alternative to either option.

Levels of information security also vary among the private, public and community deployment models, with private clouds having the most restricted user access and, most likely, the least exposure to threats.

The hybrid cloud offers a combination of two or more deployment models with varying levels of security, as needed. Users may choose to leverage private or community clouds for their most business-critical data while using the public cloud for data that are already publicly available or for other non-classified data or applications. Some enterprises may just accept the risk and choose to use the public cloud regardless of data classification.

Each enterprise must make its own choice based on its unique situation and needs. Some pertinent questions to ask to determine which deployment model may be the most appropriate include:

- Is the business process to be used in the cloud a nonstandard solution?
- What type and number of interdependencies are there among the business processes involved?
- Is there a difference between the cloud solution and a standard IT-based solution?
- Is the application custom?
- Is the hardware custom?
- Is the OS custom?
- Is the business driver cloud-compatible?

It is important to note that the offerings of CSPs are not necessarily standardized. Prospective cloud clients must use care in determining the amount of security provisioning they will require depending on the type of application and the security classifications of the data they plan to promote to the cloud.

1.6 Chapter 1 Knowledge Check

REVIEW QUESTIONS

1. One of the **MAIN** characteristics of cloud technology service models is its:

   A. On-demand service
   B. Physical device ownership
   C. Measured elasticity
   D. Limited network access

2. Cloud models provide the customer the ability to _______ risk:

   A. Accept
   B. Mitigate
   C. Transfer
   D. Ignore

3. The ability to run a guest operating system is a characteristic of:

   A. Simplification
   B. Elasticity
   C. Resource pooling
   D. Virtualization

4. The difference between a hypervisor and a container is that a container runs on top of a single OS and emulates a guest OS.

   A. True
   B. False

5. When considering a cloud vendor, special attention should be paid to:

   A. The cloud provider's culture
   B. The skill set of your internal IT team
   C. Data migration capabilities
   D. Contract usability

6. After data migration into the cloud, a _____ should be conducted.

   A. Thorough investigation
   B. Post-mortem
   C. Implementation review
   D. Post-implementation review

7. The **BEST** cloud deployment model for an on-premise single enterprise would be:

   A. Private
   B. Public
   C. Hybrid
   D. Community

8. Cloud deployment models are standardized throughout the industry.

   A. True
   B. False

Answers on page 21

Chapter 1 ANSWER KEY

Review Questions

1. **A. On-demand service. Refer to page 12.**
   B. Physical device ownership
   C. Measured elasticity
   D. Limited network access

2. A. Accept
   B. Mitigate
   C. Transfer. Refer to page 13.
   D. Ignore

3. A. Simplification
   B. Elasticity
   C. Resource pooling
   D. Virtualization. Refer to page 14.

4. A. True Refer to pages 15-16.
   B. False.

5. A. The cloud provider's culture
   B. The skill set of your internal IT team
   C. Data migration capabilities. Refer to page 17.
   D. Contract usability

6. A. Thorough investigation
   B. Post-mortem
   C. Implementation review
   D. Post-implementation review. Refer to page 17.

7. **A. Private. Refer to page 17.**
   B. Public
   C. Hybrid
   D. Community

8. **A. True. Refer to page 18.**
   B. False

Chapter 2:

Cloud Service Models

Cloud Service Models

Cloud Service Models

2.1 Learning Objectives

After completing this chapter, learners will be able to:

1. Identify cloud service models.
2. Distinguish characteristics of SaaS, PaaS and IaaS models.
3. Use service models as potential solutions for project implementation.
4. Explain solution stacks and the decisions that need to be made to implement them.

2.2 Overview

A cloud service model (e.g., cloud delivery model) specifies the capabilities offered to users and the applications supported. Many questions must be asked and answered to determine which delivery model is right for the enterprise, including:

- Are the data to be stored/processed in the cloud sensitive?
- Are the data to be stored/processed in the cloud critical?
- Is anything other than data to be stored/processed in the cloud?
- Are business processes to be performed in the cloud critical?
- Is the infrastructure offered by the cloud adequate for the purpose? Is it predictable?
- Are there any legal/compliance or regulatory impediments to hosting the data in the cloud?
- Are data ownership issues clearly defined and agreed upon?
- What jurisdictions preside over the location of the data?
- Is an adequate and appropriate service level agreement (SLA) being offered?
- Is the cloud solution cost-efficient?
- Does the CSP have a disaster recovery plan (DRP)/business continuity plan (BCP)?
- Are upgrades in service available?

There are three basic cloud delivery models that are examined in this publication (see **figure 2.1**), each offering a distinct computing service to the enterprise that uses it:

- **Software as a Service (SaaS)**—Provides a business application that is used by many individuals or enterprises concurrently
- **Platform as a Service (PaaS)**—Provides an application development sandbox in the cloud
- **Infrastructure as a Service (IaaS)**—Provides online processing or data storage capacity

Figure 2.1—Cloud Delivery Methods

Service Model	Description	Considerations
SaaS	Capability to use the provider's applications running on cloud infrastructure. The applications are accessible from various client devices through a thin client interface such as a web browser (e.g., web-based email).	SaaS provides applications that are complete and available on demand to the end customer. Traditional licensing and asset management are changed.
PaaS	Capability to deploy onto the cloud infrastructure customer-created or customer-acquired applications developed using programming languages and tools supported by the provider.	PaaS provides an application development sandbox and is specifically designed for developers.
IaaS	Capability to provision processing, storage, networks and other fundamental computing resources, offering the customer the ability to deploy and run arbitrary software, which can include OSs and applications. IaaS puts these IT operations into the hands of a third party.	IaaS can provide infrastructure services such as servers, disk space, network devices and memory and is designed for users wanting complete freedom with regard to the OS and applications they use.

Source: ISACA, *Controls and Assurance in the Cloud: Using COBIT® 5*, USA, 2014, *https://www.isaca.org/bookstore/cobit-5/cb5ca*

Before examining these three in further detail, it is worth noting that they are not the only delivery models available. Cloud service offerings are constantly undergoing major changes and evolving into new and better service offerings. Within these three "traditional" service deliveries, a variety of new services are emerging that are closely related. Most of the new services focus on replacing traditional in-house IT services with cloud variants and on the activities with which internal IT departments frequently struggle. These services benefit an enterprise in multiple ways:

- Easier access to new technologies; smaller enterprises gain access to a wide range of solutions that were previously beyond their financial reach
- Technical knowledge of the CSP and access to newer technology that can provide a competitive edge through the services delivered
- Correct and structured implementation of the targeted services

The adoption of cloud technologies should always be subject to careful study and should be aligned and integrated with the internal processes and procedures of an enterprise. The following "niche" services are increasingly used in the marketplace:

- **Security as a service (SecaaS)**[3]—SecaaS comes in two major forms:
  - The CSP provides standalone managed security services ranging from antivirus scanning and mail security to full deployments of end-point security.
  - The CSP offloads appliance utilization for the client, and CPU- and memory-intensive activities are moved to cloud services.
- **Disaster recovery as a service (DRaaS)**—The CSP offers its cloud infrastructure to provide an enterprise with a disaster recovery (DR) solution. In most cases, the CSP not only provides backup equipment and storage, but also provides services for a BCP, if it is not yet available. This significantly reduces the cost for an in-house DR infrastructure. In addition, offsite storage means that the DR environment is less likely to fail in the case of a major disaster.
- **Identity as a service (IDaaS)**—IDaaS currently has two interpretations:
  - The management of identities in the cloud that is separated from the users and applications that use the identities. This can be either managed identity services, including provisioning, or management for both

[3] The Cloud Security Alliance® (CSA), a nonprofit organization with a mission to promote best practices for providing security assurance with cloud computing, is a valuable source of information on this topic (*cloudsecurityalliance.org*).

onsite and offsite services. Delivering a single sign-on (SSO) solution can also be part of the cloud service offering.
  - The delivery of an Identity and Access Management (IAM) solution. IDaaS is often a hybrid solution (a federated model) in which access and roles are configured by the CSP and users are authorized by enterprise internal solutions.

- **Data storage and data analytics as a service (big data)**—Big data makes it possible to analyze all types of data by taking away the constraints on volume, variety, velocity and veracity. These constraints are removed through a synergy between new technologies and the extended capabilities provided by cloud computing. Limitless volume availability and variety allow enterprises to reuse their "old" data for new purposes. Furthermore, big data technology facilitates the ability of enterprises to find patterns in their current data, which influences their way of doing business.
- **Information as a service (InfoaaS)**—This service builds on the big data concept. Rather than providing the raw data or algorithms that are used for trending, InfoaaS provides the required information. With this service, the result of a query is more important than the query itself.
- **Integration platform as a service (IPaaS)**—IPaaS, which is also called "cloud integrator," is defined as "a set of automated tools for connecting software applications that are deployed in different environments...often used by large business-to-business (B2B) enterprises that need to integrate on-premises applications and data with cloud applications and data."[4] Many enterprises are implementing a hybrid model in which some of their data, applications, services and infrastructure are maintained locally onsite, while others are provisioned by a cloud provider. Cloud integrators can help address the complexity of integrating these business resources without the need to constantly modify and maintain diverse and often incompatible applications.
- **Forensics as a service (FRaaS)**—FRaaS is a multi-tenant capacity framework that enables forensic analysis on digital evidence to be provided as a service in the cloud.[5]
- **Blockchain as a service (BaaS)**—BaaS utilizes cloud technology to deploy and support blockchain systems.

Because the marketplace and the cloud environment are so dynamic, other targeted deployment models are undoubtedly already in development or in actual use.

2.3 Software as a Service (SaaS)

SaaS is a method of software delivery and licensing in which software is accessed online via a subscription, rather than bought and installed on individual computers onsite, and is used by many individuals or enterprises concurrently. The applications are accessible from various client devices through a thin client interface, such as a web browser. SaaS provides the most-used cloud applications to the vast majority of online users. This model also provides the customer with the least amount of control as it is mostly managed by the CSP. Examples of consumer-directed SaaS applications include Gmail and Netflix. Office 365 and QuickBooks are common business-directed SaaS applications.

SaaS can be modified for specific business uses and individual users. An enterprise can customize the user interface (UI) to change the look and feel of the program, and modify specific areas, such as data fields, to alter what data appear. Several business process features can also be turned off and on at will. In addition, users can often edit their own personal workspace, such as a dashboard or task list. When the workspace is edited, it shows only the information they need to see for their unique work style.

[4] TechTarget, "iPaaS (integration platform as a service)," *searchcloudcomputing.techtarget.com/definition/iPaaS-integration-platform-as-a-service*

[5] Soni, M.; M.K. Barti; "FraaS: A Framework for Digital Forensic Services in a Cloud-based Environment," *The International Journal of Forensic Science*, 2015, 1, p. 15-22, *ijofcs.org/V10N1-PP02-FraaS.pdf*

Despite SaaS's utility, it is not necessarily perfect in all situations. Two areas in which SaaS may not be the right solution are:

- Applications that require fast processing of real-time data
- Applications that are prohibited from hosting data externally because of legislation or other regulations[6]

2.4 Platform as a Service (PaaS)

A computing platform or digital platform is the environment in which a piece of software is executed (run). PaaS provides the capability to deploy customer-created, or -acquired, applications that are developed using programming languages and tools that are offered by the provider. Stated briefly, it provides an application development sandbox in the cloud. Examples of PaaS include Microsoft Azure and the Google App Engine.

With PaaS, the CSP offers enterprise developers elemental Service-Oriented Architecture (SOA) application building blocks to configure a new business application. This may include such underlying cloud infrastructure components as servers, networking equipment, operating systems, storage services, middleware and databases.[7] In-house development requires development, testing and user acceptance platforms, all separate from the production environment. Through PaaS, enterprise developers can rent their development environment (hardware and software), complete with an SOA tool kit, and the enterprise is charged for only the time the tools and environment are in use. The CSP owns and is responsible for operating, configuring and maintaining its technology offerings.

While PaaS has many attractive features, it also is not the perfect solution for all needs. Examples where PaaS is likely to fall short include situations in which:

- Applications need to be highly portable relative to hosting
- Proprietary language would impact development or hinder later moves to another provider
- Application performance requires that the underlying hardware and software be customized[8]

2.5 Infrastructure as a Service (IaaS)

IaaS is a cloud service that provides online processing or data storage capacity. It can be considered a virtual data center. This cloud service is ideal for enterprises considering very large, one-time processing projects or infrequent, extremely large data storage requirements, such as test environments. IaaS offers the capability to provision processing, storage, networks and other fundamental computing resources, enabling the customer to deploy and run arbitrary software, which can include OSs and applications. Other uses for IaaS are handling unpredictable demand and increasing storage needs, simplifying planning and management of backup and recovery systems, and supporting web applications. An example of IaaS is Amazon Web Services (AWS), which provides on-demand computing power, storage and networking capabilities.

Like SaaS and PaaS, IaaS is not always the ideal solution. For example, situations in which regulatory compliance would make offshoring or outsourcing data storage and processing difficult are likely to prove problematic in an IaaS environment.[9]

[6] Rackspace Support, "Understanding the cloud computing stack: SaaS, PaaS, and IaaS," 1 April 2020, *support.rackspace.com/how-to/understanding-the-cloud-computing-stack-saas-paas-iaas/*

[7] Violino, B.; "What is PaaS? Platform-as-a-service explained," *InfoWorld*, 19 July 2019, *www.infoworld.com/article/3223434/what-is-paas-software-development-in-the-cloud.html*

[8] *Op cit* Rackspace Support

[9] *Ibid.*

Figure 2.2 illustrates the three cloud delivery models and three cloud deployment models, as they address cloud trust issues.

Figure 2.2—Cloud Computing Service Delivery and Deployment Model

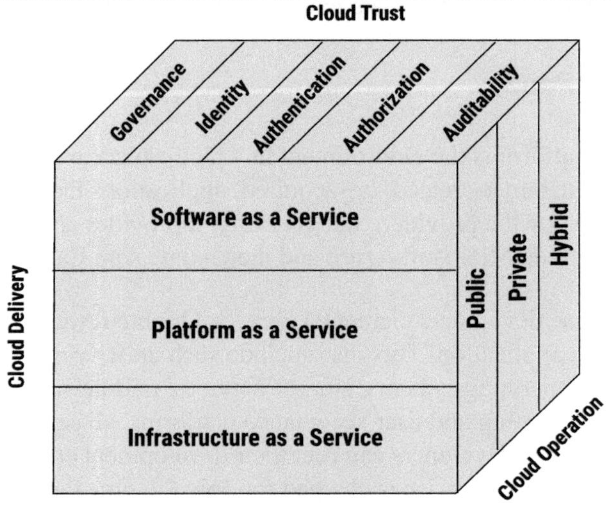

Source: Cloud Security Alliance®, Trusted Cloud Initiative Reference Architecture. ©Cloud Security Alliance, *https://downloads.cloudsecurityalliance.org/initiatives/tci/TCI_Reference_Architecture_v2.0.pdf*.
Used with permission.

2.6 Solution Stacks

A solution stack is a set of different programs or application software that are bundled together to produce a desired result or solution. This may refer to any collection of unrelated applications taken from various subcomponents working in sequence to present a reliable and fully functioning software solution.[10] A widely known example of a solution stack is Microsoft Office.

There are many types of stacks; web, software/application, virtualization, server and storage are the most common. Cloud computing itself, with its many "as a service" offerings, is often described as a stack, due to its wide range of services built on top of each other under the name of the cloud. Early in its development, six layers of the cloud computing stack were identified (listed from the top of the stack—in terms of visibility—to the bottom):[11]

- **Clients**—Computer hardware and/or software that rely on the cloud for application delivery or that is specifically designed for delivery of cloud services
- **Services**—Software systems designed to support machine-to-machine interaction over a network that can be accessed by other cloud computing components, software or end users
- **Applications**—Leveraging the cloud in software architecture, often enabling users to avoid installing and running the application on their own computers
- **Platform**—Delivery of a platform or solution stack (as a service); PaaS
- **Storage**—Delivery of data storage as a service
- **Infrastructure**—Delivery of computer infrastructure, often a platform virtualization environment, as a service; IaaS

[10] Techopedia, "Solution Stack," *www.techopedia.com/definition/28154/solution-stack*

[11] Johnston, S.; "Taxonomy: The 6 layer Cloud Computing Stack," Sam Johnston Scribes, 17 September 2008, *samj.net/2008/09/17/taxonomy-the-6-layer-cloud-computing-stack/*

Over time, however, this six-layer view of the cloud computing stack has been streamlined and simplified, most commonly to the generally accepted three-layer view, from top to bottom: SaaS, PaaS, IaaS.

Enterprises that use the cloud are likely to have little knowledge or "visibility" into the people, processes and technology supporting their own information assets. The lack of visibility is also known as "abstraction;" to counter the effects, the CSP should provide full details to customers on how their assets are managed. The level of abstraction or visibility provided by the CSP becomes extremely important when evaluating risk. In fact, each service model corresponds to an abstraction level based on the number of layers in the Internet Protocol (IP) stack being replaced by the cloud. For this reason, IaaS represents the lowest abstraction level (infrastructure only) and SaaS the highest (application + middleware + infrastructure). The higher the abstraction level, the higher the risk or the number of threats to take into account because risk is cumulative (**figure 2.3**).

However, CSPs often offer only visibility into the cloud stack corresponding to the service model chosen. Security professionals must be aware of this factor when evaluating a move to the cloud. A common mistake is to assume that SaaS will not also be subject to risk related to infrastructure; however, risk and threats are there. They are on a layer that is less visible because it is no longer under the operational responsibility of the enterprise, but the CSP instead.

Figure 2.3—Stacks in Cloud Service Models

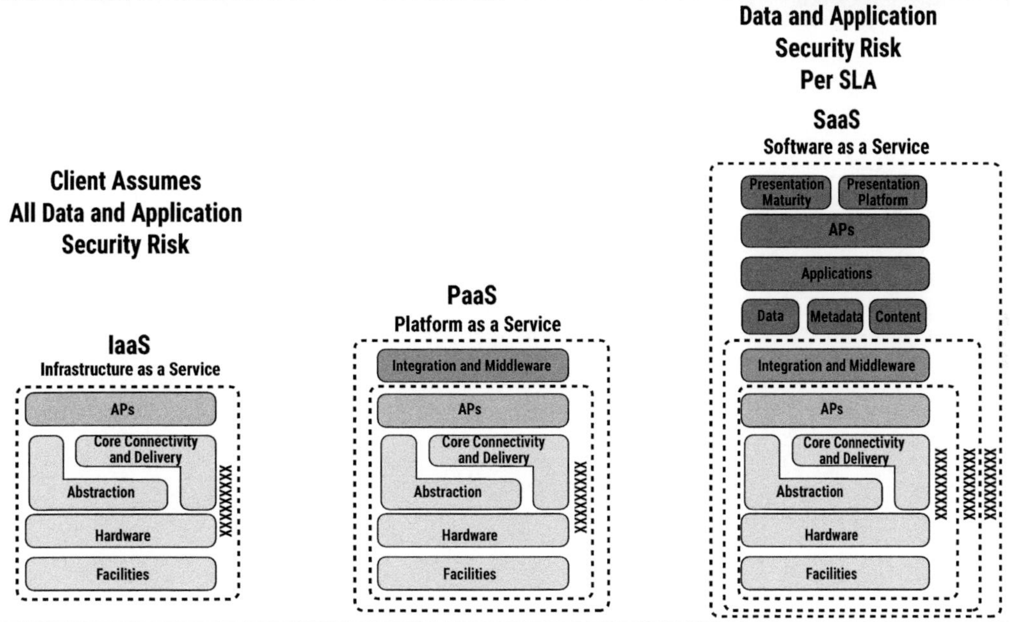

Source: Cloud Security Alliance®, *Universal Model*. ©Cloud Security Alliance. Used with permission.

Page intentionally left blank

2.7 Chapter 2 Knowledge Check

REVIEW QUESTIONS

1. Office 365® is an example of:

   A. IaaS
   B. PaaS
   C. OaaS
   D. SaaS

2. For which purpose might SaaS not be the **BEST** option:

   A. Specific business uses
   B. Email
   C. Fast real-time data processing
   D. Video streaming services

3. The PaaS model would be **BEST** used for implementing:

   A. Communication platforms
   B. Full virtualized infrastructure
   C. File storage
   D. Programming development environments

4. PaaS would be a great solution for a proprietary piece of software that requires massive amounts of system resources (RAM, CPU, etc.).

   A. True
   B. False

5. The IaaS model would be **BEST** used for:

   A. Subscription services
   B. Large data storage requirements
   C. Single operating system instances
   D. VPN implementation

6. AWS®, Google Cloud™ and Microsoft® Azure® are all examples of IaaS:

   A. True
   B. False

7. Which of the following is not a layer of cloud computing solution stack:

   A. Services
   B. Networks
   C. Application
   D. Storage

Answers on page 32

Chapter 2 ANSWER KEY

Review Questions

1. A. IaaS
   B. PaaS
   C. OaaS
   D. SaaS. Refer to page 24.

2. A. Specific business uses
   B. Email
   C. Fast real-time data processing. Refer to page 27.
   D. Video streaming services

3. A. Communication platforms
   B. Full virtualized infrastructure
   C. File storage
   D. Programming development environments. Refer to page 27.

4. A. True
   B. False. Refer to page 27.

5. A. Subscription services
   B. Large data storage requirements. Refer to page 27.
   C. Single operating system instances
   D. VPN implementation

6. **A. True. Refer to page 27.**
   B. False

7. A. Services
   B. Networks. Refer to page 28.
   C. Application
   D. Storage

Chapter 3:

Cloud Governance

Cloud Governance

Cloud Governance

3.1 Learning Objectives

After completing this chapter, learners will be able to:

1. Identify the challenges of governance in the cloud.
2. Explain the reasons for implementing and maintaining cloud governance.
3. Articulate the risk involved with cloud implementation.
4. Describe cloud vendor selection processes.
5. Define the need for portability of services in a service level agreement (SLA).

3.2 Overview

Governance is the method an enterprise adopts to evaluate stakeholder needs, conditions and options to ensure that balanced, agreed-on enterprise objectives are achieved. It involves setting direction through prioritization and decision making, and monitoring performance and compliance against agreed-on direction and objectives. Governance in the cloud takes those objectives and activities and focuses them on the cloud environment—especially important in many enterprises, given the strategic contribution of IT and cloud computing toward achieving business goals.

A move to the cloud will cause enterprises to make changes internally to help ensure that they continue to meet performance objectives, their technology provisioning and business are strategically aligned, and risk is managed. For example, the role of the chief information officer (CIO) is transitioning from managing operations to managing IT as a service value chain. Previously, the CIO may have run the internal "IT factory"; now, both enterprises and external service providers have become producers and consumers of services. With cloud computing, the CIO must weave together and optimize this value chain to best support various customers and enable an enterprise's business.

Ensuring that IT is aligned with the business, systems are secure and risk is managed is challenging in any environment and even more complex in a third-party relationship. As with other organizational changes, adoption of cloud computing will necessitate some adjustments to the way business processes are handled. Business processes such as data processing, development and information retrieval are examples of potential change areas. Additionally, processes detailing the way information is stored, archived and backed up will need revisiting.

For enterprises to benefit from cloud computing, a clear governance strategy and management plan must be developed. The governance strategy should set the direction and objectives for cloud computing within the enterprise; the management plan should execute the achievement of the objectives. Cloud computing governance creates business-driven policies and principles that establish the appropriate degree of investments and control around the life-cycle process for cloud computing services.

Just a few reasons enterprises should implement and maintain sound cloud governance are the organization's need to:[12]

- Manage increasing risk effectively, including security, compliance, privacy, projects and partners.
- Ensure continuity of critical business processes that now extend beyond the data center.
- Communicate clear enterprise objectives internally and to third parties.

[12] Stroud, R.; "Providing Governance in a Rapidly Changing World," ISACA Euro Computer Audit, Control and Security (EuroCACS) conference 2010, Budapest, Hungary

- Adapt easily. Flexibility, scalability and services are changed in the cloud, enabling the enterprise and business practices to adjust, create new opportunities and reduce costs.
- Facilitate continuity of IT knowledge, which is essential to sustain and grow the business.
- Handle a myriad of regulations.

Governance starts at the top of the organization and, in fact, is unlikely to achieve its mission without complete buy-in from the board of directors. Establishing a clear direction that is aligned with enterprise strategy calls on members of the board to develop a clear understanding of cloud computing benefits and how to maximize them through effective end-to-end governance practices. This requires the board to see cloud computing not as an IT project, but rather as a business technology strategy. This understanding helps ensure that stakeholder needs are considered and met while risk and resource utilization are optimized.

IT, with its particular requirements and terminology, has historically been viewed as a cost center rather than a corporate asset. However, the cloud presents the opportunity to fully align IT with the goals and objectives of the enterprise as a whole. To "sit at the table" of the enterprise's governance programs, IT must adapt to methodologies and the language used in other areas of the enterprise's governance.

3.3 Business Drivers to Cloud Solutions

Cloud computing is viewed as a significant change to the platform in which business services are translated, used and managed. Many consider it to be as large a shift in IT as was the advent of the personal computer or Internet access. However, the rollout of those earlier technologies encompassed a slower development phase, whereas cloud technologies and solutions have come together more rapidly for implementation. This means enterprise decision makers must quickly develop an understanding of the business drivers that may be addressed by adoption of a cloud solution, as outlined in **figure 3.1**.

Figure 3.1—Business Drivers and Cloud Response

Business Driver	Description	Cloud Response
Optimized resource utilization	Organizations are best served by having just the resources they need when they need them. Purchasing hardware and software "just in case" results in extra expense and storage space.	• By using a pay-as-you-go cloud solution, resources become available when needed and are liberated when no longer needed; there is near-to-perfect alignment with actual demand.
Cost savings	Enterprises seek to reduce IT expenses and restructure these expenses to spread them out over time.	• Increased server utilization plus the transition of computational capability from acquired and maintained computers to rented cloud services change the computing cost paradigm from a capital expenditure (CAPEX) to an operational expenditure (OPEX), with potentially significant up-front and total cost savings. • Flexible, on-demand services enable solution testing without significant capital investments and provide transparency of usage charges to drive behavioral change within organizations. • The number of workstations in the office can be reduced and some employees can be allowed to work from home to save costs further.

Figure 3.1—Business Drivers and Cloud Response *(cont.)*		
Business Driver	**Description**	**Cloud Response**
Better responsiveness	In a fast-paced business environment, the ability to respond to developments as they occur provides competitive advantage.	• The cloud offers on-demand, agile, scalable and flexible services that can be implemented quickly, which provide organizations with the ability to respond to changing requirements and peak periods. • Enterprise operations can be monitored effectively, even in real time, enabling improved responsiveness to evolving situations and opportunities.
Faster cycle of innovation	Success in a competitive environment calls for faster development of products and services.	• Cloud computing makes patch management and upgrades to new versions more flexible. • Cloud users can upgrade to a new software version often by doing nothing more than typing a different URL into the web browser. • Staff members can easily share data and collaborate to complete projects even from different locations. Field workers can easily share real-time data and updates with those in the office. • Cloud computing eliminates redundant or repetitive tasks such as data re-entry and scalability.
Reduced time for implementation	New business initiatives brought online in a traditional IT enterprise may require weeks or months to complete (and accrue CAPEX).	• Cloud computing provides processing power and data storage as needed and at the capacity needed, in near-real time.
Resilience	Lack of resilience can increase the potential for a costly, time-consuming and potentially reputation-damaging system failure.	• The failure of one component of a cloud-based system has less impact on overall service availability and reduces the risk of downtime. • By storing data in the cloud, which can be safer than storage on physical servers and in data centers, enterprise data will always be available if users have an Internet connection. • If the enterprise's on-premise data security is compromised, perhaps through the theft of laptops or tablets, the enterprise that has data stored in the cloud can delete confidential information remotely or move it to a different account.

Each cloud opportunity or program has many variables, benefits and risk that affect the decision of whether a cloud application should be adopted from a risk/business value standpoint. The enterprise must weigh those variables to decide whether the cloud is an appropriate solution.

The following questions can help identify the strategic value that cloud services may provide to the enterprise and the impact that cloud could have on enterprise resources and controls. This, in turn, will help identify the cloud deployment required, which will lead to an understanding of the type of CSP needed. Upper management (C-suite) must be prepared to answer these questions in terms that can be easily related to the business benefits cloud computing will provide:

1. Do management teams have a plan for cloud computing? Have they weighed value and opportunity costs? The risk of cloud adoption may be inconsequential when compared to the lost opportunity to transform the enterprise with effective and strategic use of cloud computing—especially if competitors take steps to leverage those same opportunities. From a strategic perspective, cloud computing can be a vehicle to:
   - Gain competitive advantage
   - Reach new markets

- Improve existing products and services
- Retain existing customers
- Increase productivity
- Contain cost
- Develop products or services that could not be possible without cloud services
- Break geographic barriers

2. How do current cloud plans support the enterprise's mission? Cloud services should support efforts to achieve business objectives, which are derived from stakeholder needs (as vetted by the leadership team). Cloud initiatives should have a clear and traceable link to the enterprise strategy so that the value expected from cloud services is clearly defined, accepted and measurable. This link also helps determine the priority assigned to cloud initiatives and supports the development of metrics to measure results against expectations.

   Alignment between cloud objectives and enterprise objectives is critical for effective risk management and cost containment. The potential benefits of cloud services can be enticing, but with reward comes risk. The enterprise must decide whether the potential risk is within acceptable limits.

3. Have executive teams systematically evaluated organizational readiness? Pressure points result when:
   - Cloud computing implementations conflict with enterprise culture
   - Skills that are required to support cloud solutions are not available
   - Cloud-related processes conflict with other established processes
   - Organizational structure does not maximize cloud effectiveness or efficiency

   Evaluating the readiness of the enterprise in anticipation of the adoption of cloud services avoids the need for after-the-fact culture, skill or process changes to remove unanticipated pressure points. A systematic readiness assessment can help management identify additional cost and risk that should be factored into the decision-making process. This readiness assessment should include the following:
   - **Policies and procedures**—New policies and procedures that guide the adoption, management and proper use of cloud computing may be needed.
   - **Processes**—Existing processes using traditional IT services may need to be re-engineered to incorporate new activities that are related to using cloud services.
   - **Organizational structures**—Cloud management may require new organizational capabilities or modifications to existing organizational structures, particularly in IT operations and support.
   - **Culture and behavior**—Organizational culture and behavior can be critical for the successful adoption of cloud solutions.
   - **Skills and competencies**—Procurement, legal, compliance and audit are some examples of functions that may need to develop necessary skills to manage cloud services from evaluation and sourcing to operations and retirement.

4. Have management teams considered what existing investments might be lost in their cloud planning? Cloud computing may not be an immediate and clean fit with the existing technology portfolio of the enterprise. The adoption of a cloud service may, for example, obviate technology investments already made that have not reached their planned end date. The decision about when and how to realize that loss must be considered carefully. Areas to consider include:
   - **Processes**—The IT organization may need to adapt processes such as sourcing and change management.
   - **Culture and behavior**—Cloud services may demand faster turnaround from the IT organization, which may necessitate changes in internal processes and tools.
   - **Services, infrastructures and applications**—The enterprise may need to update data centers, software applications and network infrastructures, which may result in some level of lost investment being realized.

- **Skills and competencies**—The IT organization will need to either develop or acquire the skills required to support users of cloud services if those skills do not already exist within current staffing.

5. Do management teams have strategies to measure and track the value of cloud return vs. risk? Before deciding to adopt cloud computing, the board should give management teams the task of ensuring that proper reporting mechanisms are in place to measure value and risk aligned with enterprise goals.

The enterprise should also consider stakeholders' potential issues when evaluating business drivers for a move to the cloud. Do stakeholders have needs that are beyond the capabilities of the current IT systems? Have stakeholders identified business opportunities that require the support of cloud applications? Do stakeholders anticipate business gains that can be realized only through cloud usage? Are stakeholders concerned about specific compliance issues in the cloud?

Ultimately, cloud is not necessarily the right solution for every organizational need. The most common technical reason enterprises find to not move to the cloud is that the cost of customization outweighs the benefits of the cloud solution. The type of cloud service selected is critical, as is how it is managed. Thinking strategically about benefits, costs and risk is paramount and must be done upfront, before any contract is signed.

3.4 Risk Associated With Cloud Solutions

Just as the benefits associated with the cloud vary based on multiple factors related to the enterprise and the technology, so too does the associated risk. Among the factors that can impact cloud-related risk are:

- **Type of cloud service model**—Each cloud service model (SaaS, PaaS, IaaS, etc.) has varied business purposes and levels of business risk.
- **Robustness of the enterprise's existing IT operations**—Enterprises' current governance, risk management and information security enablers must be well-defined and managed within the existing IT operations. New threats and vulnerabilities may be identified in the cloud, but if the enterprise is prepared and its current IT operations are able to handle these issues, the overall risk to the enterprise may be lower.
- **Current level of business risk acceptance**—The level of risk an enterprise is willing to accept varies among industries and among enterprises within the same industry.
- **Aggregated "street value" of the data to be promoted to the cloud**—Enterprises need to assess the value of the data promoted to the cloud in terms of how tempting the data may be to those with malicious intent.
- **Internal security classification of data being promoted to the cloud**—As with "street value," data have internal value to the enterprise, which provides the enterprise with a vested interest in keeping them proprietary and not releasing data publicly.
- **Identified compliance obligations of the data shared within the cloud**—Personally identifiable information (PII) security controls and financial reporting compliance are two prime examples of compliance obligations that need to be managed in the cloud.
- **CSP risk**—In the absence of commonly accepted cloud security standards, CSPs may take different approaches to cloud security. They should follow best practices and use internationally accepted standards for general information security. Enterprises should exercise due diligence in evaluating and selecting a CSP; having their own well-defined requirements will enable enterprises to reap maximum benefits.

Information assets such as those placed in the cloud can be roughly categorized as data, applications and processes. These assets are commonly subject to the following risk events:

- **Unavailability**—The asset is unavailable and cannot be used or accessed by the enterprise. The cause can be accidental (failure of the infrastructure), intentional (distributed denial-of-service [DDoS] attacks) or legal (subpoena of database holding all data in a case of multi-tenancy architecture where one client's data are subject to legal investigation).

- **Loss**—The asset is lost or destroyed. The cause can be accidental (natural disaster, wrongful manipulation, etc.) or intentional (deliberate destruction of data).
- **Theft**—The asset has been intentionally stolen and is now in possession of another individual/enterprise. Theft is a deliberate action that can involve data loss.
- **Disclosure**—The asset has been released to unauthorized staff/enterprises/organizations or to the public. Disclosure can be accidental or deliberate. This also includes the undesired, but legal, access to data due to different regulations across international borders.

Data are commonly the most valuable assets and the most probable targets of attacks in the cloud. However, the risk related to applications and processes should not be overlooked. The business impact of long DDoS attacks cannot always be absorbed by an enterprise; although no data loss or disclosure is suffered, paralyzing the systems has direct, negative effects on the data. Disclosure of details about how applications handle critical information or about internal enterprise processes could pave the way to more selective attacks with greater consequences.

Figure 3.2 describes at a high level the possible impact of the four risk events on assets.

<table>
<tr><th colspan="5">Figure 3.2—Impact of Risk Events on Information Assets</th></tr>
<tr><th>Type</th><th>Unavailability</th><th>Loss</th><th>Theft</th><th>Disclosure</th></tr>
<tr><td>Data</td><td>• Disruption of activities
• Lack of resources to maintain business as usual
• Possibility of data poisoning</td><td rowspan="2">• Disruption of activities
• Required activation of backup restore procedures (DRP)
• Possibility of partial loss of the asset (depending on the recovery point objective (RPO)
• Financial loss associated with recovery efforts</td><td>• Business competitive disadvantage
• Possibility of blackmail
• Loss of credibility with customers/clients</td><td>• Damage to company reputation or image
• Possibility of regulatory sanctions
• Financial impact</td></tr>
<tr><td>Applications/ processes</td><td>• Disruption of activities
• Lack of resources to maintain business as usual</td><td colspan="2">• Higher risk/threat of more selective attacks to data</td></tr>
<tr><td colspan="5">Source: ISACA, Security Considerations for Cloud Computing, USA, 2012, https://www.isaca.org/bookstore/audit-control-and-security-essentials/scc</td></tr>
</table>

Given the multiple and diverse potential areas of risk in the cloud, it is impossible to accumulate an exhaustive list of them all. However, certain risk issues are common to many enterprises and CSP environments. One of the most pressing is information security, which merits a closer look, but other areas of risk can be equally concerning.

3.4.1 Security

Security and privacy must be considered from the moment management considers cloud computing a possibility. Security and privacy requirements should be part of the due diligence, business case studies and vendor selection. Whenever possible, the enterprise should evaluate the provider's control environment to ensure that the enterprise's security and privacy needs will be met satisfactorily. When a close assessment is not possible, the enterprise must request proof that the provider has implemented the necessary controls to protect the enterprise's assets.

Cloud services aggregate data from thousands of businesses. Each new client, bringing a new store of data, increases the value of that CSP as a potential target. This concentrates risk on a single point of failure. A disaster at a cloud provider can affect every one of its customers. A CSP's failure to apply, maintain and monitor the appropriate security measures can result in the commingling of data among clients, compromised data, stolen data and incomplete data deletion. Data breaches can result in diminished customer trust, revenue loss and loss of intellectual property for both the client enterprise and the CSP.

Enterprises may wish to consider securing a Service Organization Control 3 (SOC 3) report, which outlines information related to a service organization's internal controls for security, availability, processing integrity, confidentiality or privacy. Because they are general use reports, SOC 3 reports can be freely distributed. The five areas covered in SOC 3 reports are the focuses of the American Institute of Certified Public Accountants (AICPA) Trust Services Principles and Criteria:

- **Security**—Information and systems are protected against unauthorized access, unauthorized disclosure of information and damage to systems that could compromise the availability, integrity, confidentiality and privacy of information or systems, and affect the entity's ability to achieve its objectives. Security refers to the protection of information during its collection or creation, use, processing, transmission and storage, as well as the systems that use electronic information to process, transmit or transfer, and store information to enable the entity to meet its objectives. Controls over security prevent or detect the breakdown and circumvention of segregation of duties; system failure; incorrect processing; theft or other unauthorized removal of information or system resources; misuse of software; and improper access to or use of, alteration, destruction, or disclosure of information.
- **Availability**—Information and systems are available for operation and use to meet the entity's objectives. Availability refers to the accessibility of information used by the entity's systems as well as the products or services provided to its customers.
- **Processing integrity** (over the provision of services or the production, manufacturing, or distribution of goods)—Processing integrity refers to the completeness, validity, accuracy, timeliness and authorization of system processing. Processing integrity addresses whether systems achieve the aim or purpose for which they exist and whether they perform their intended functions unimpaired, free from error, delay, omission, and unauthorized or inadvertent manipulation.
- **Confidentiality**—Confidentiality addresses the entity's ability to protect information (all types of sensitive data, including personal information) designated as confidential from its collection or creation through its final disposition and removal from the entity's control in accordance with management's objectives. Information is confidential if the custodian (for example, an entity that holds or stores information) of the information is required to limit its access, use and retention, and restrict its disclosure to defined parties (including those who may otherwise have authorized access within its system boundaries). Confidentiality requirements may be contained in laws or regulations or in contracts or agreements that contain commitments made to customers or others.
- **Privacy**—Personal information is collected, used, retained, disclosed and disposed of to meet the entity's objectives. Although confidentiality applies to various types of sensitive information, privacy applies only to personal information. The privacy criteria include notice and communication of objectives; choice and consent; collection; use, retention and disposal; access; disclosure and notification; quality; and monitoring and enforcement.[13] Industry PII regulations such as HIPAA, CCPA and PCI-DSS should also be a consideration for privacy.

Other standards, certifications and frameworks are available, including useful material from ISACA and the Cloud Security Alliance.

It is very important for both the enterprise and the CSP to understand that cloud security is the responsibility of all parties involved in the service contract. Depending upon the deployment model and SLAs, the provider may be responsible for implementing and maintaining the necessary controls to protect its clients' assets. Likewise, the

[13] American Institute of Certified Public Accountants, "2017 Trust Services Criteria for Security, Availability, Processing Integrity, Confidentiality, and Privacy," March 2020, p. 6-8, *www.aicpa.org/content/dam/aicpa/interestareas/frc/assuranceadvisoryservices/downloadabledocuments/trust-services-criteria.pdf*

enterprise may be responsible for ensuring that the security and privacy controls are implemented correctly, operate as intended and meet the enterprise's requirements throughout the life of the contract. In some cases, the responsibility to protect information assets can be transferred to the provider, but accountability for the protection of information assets remains with the enterprise.

Cloud access security brokers (CASBs) are tools that the enterprise may wish to investigate relative to cloud security. They are on-premises or cloud-based software or appliances that are positioned between the enterprise's technology infrastructure and the CSP.

As with any product, functionality varies by vendor, but as a baseline, most CASBs enable the enterprise to gain visibility into both authorized and unauthorized cloud application usage throughout the organization, provide a way to discover potential regulatory noncompliance, ensure secure storage of data in the cloud, and offer a degree of threat protection that mitigates to an acceptable security risk level.

Overall, most enterprises use CASBs for three primary purposes:

- **Discovery and analysis of cloud applications**—Most organizations have a significant number of applications, tools or services running in the cloud, most of which are authorized; that is, they are tested, approved, implemented and monitored by the IT department. However, in some cases, employees need enhanced functionality in a hurry and leverage the relatively low cost of and easy access to cloud applications without going through the IT department's approval process. These instances are referred to as "shadow IT." While shadow IT can be beneficial, it also has the potential to expose the organization to significant data leakage and, potentially, a breach, which may result in financial loss, reputation damage and/or other threats such as noncompliance. Many organizations rely on their CASB to discover and catalog all the cloud services being used internally so the risk can be assessed and usage patterns identified that may help them address security issues.
- **Governance and protection of data**—Just as most organizations have an overarching business strategy, they also have a data governance strategy. Among the tactics to support the strategy are policies relating to technology and data, such as security policies, data use policies and data loss prevention (DLP) policies. Communicating awareness of policies and enforcing their compliance pose challenges even on the organization's premises; it becomes even more difficult when dealing with an off-premises third party, such as a CSP. Some organizations use their CASB to extend and enforce the pertinent activity and data security policies to the cloud, to ensure that internal business rules and procedures are applied to cloud resources (files and applications). Most CASBs provide dashboards of pertinent data that enable effective monitoring of risk.

  In addition to support with governing data, CASBs can help protect the data. For example, organizations often integrate their CASB with their authentication solution to set up adaptive access controls that are informed by intelligence generated by the CASB. This enables the CASB to perceive that an employee is trying to access highly sensitive data and to apply appropriate, adapted responses, such as blocking access or sharing ability.
- **Threat detection and incident response**—While valuable, policies are only as good as the actions taken to "walk the walk." Enterprises are looking to CASBs to take a variety of actions that will enable them to identify anomalies in data movement between on-premises and cloud applications and between cloud services—anomalies that may provide early warning of malware, a failure of encryption, or an impending loss or exposure of cloud data. The activities CASBs can support in this area cover the life cycle of detecting and responding to threats—from setting threat detection thresholds, to monitoring all cloud accounts for policy violations or developing threats, to investigating incidents and generating detailed reports. The report and logs created by the CASB—which often cover the type of threat, its causes and responses taken—can also support internal audit reviews and findings, especially when internal audit has the opportunity to help develop the report format.

The CASB is a tool, not the solution; people, capabilities, resources, processing, monitoring and response plans are required to realize the CASB's full potential.

3.4.2 Other Risk Factors

Security is not the only risk factor related to cloud computing. Other areas of concern that merit attention include:

- **Privacy**—Individuals and groups have a strong drive to protect their identity in an online environment. CSPs are responsible for taking measures to protect sensitive personally identifiable data belonging to their clients. Privacy is included in a SOC 3 report, as noted previously, but enterprises that choose not to use such a report must still consider privacy issues.
- **Compliance violations**—Moving data to the cloud does exempt an enterprise from regulations surrounding the protection and use of data. However, the involvement of a CSP in data management may make compliance more difficult.
- **Loss of control**—Transitioning assets/operations to the cloud entails the enterprise losing a degree of visibility and control over those assets/operations. Enterprises must adopt and exercise monitoring and analysis of information about their applications, services, data and users. The ease of adding cloud services can also contribute to a risk-laden loss of control in that employees can provision additional services without IT's review or consent. This can lead to a lack of alignment among in-house applications, introduction of malware, redundancies and cost overruns.
- **Vendor lock-in**—Sometimes a contract with a provider does not easily protect the enterprise from changes to services and products such as supported programming languages or development tools. This becomes an issue when an organization considers moving its assets/operations from one CSP to another (or if the CSP goes out of business). The organization discovers the cost/effort/schedule time necessary for the move is much higher than initially considered, due to factors such as nonstandard data formats, nonstandard application programming interfaces (APIs), and reliance on one CSP's proprietary tools and unique APIs.

The type and degree of risk vary according to the cloud service model and deployment model selected by the enterprise, as illustrated in **figure 3.3**.

Figure 3.3—Cloud Risk by Service and Deployment Models

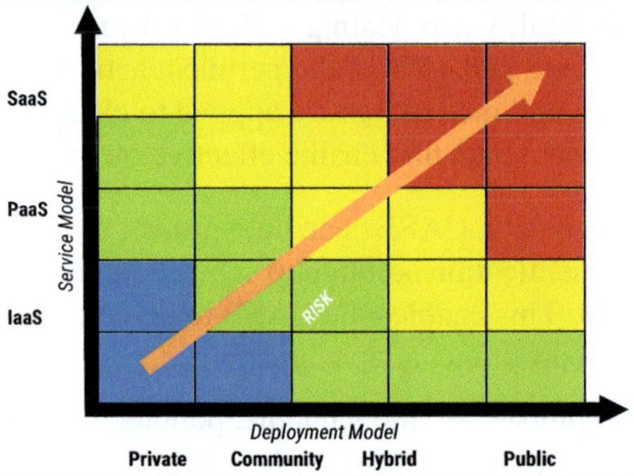

Source: ISACA, *Controls and Assurance in the Cloud: Using COBIT® 5*, USA, 2014, figure 3, *https://www.isaca.org/bookstore/cobit-5/cb5ca*

This section discusses just a few areas of cloud-related risk, most of which relate to the IT that supports the service. It is helpful to consider these multiple threats in IT risk categories and keep in mind some pointed questions that should be asked when considering a move to the cloud (**figure 3.4**).

Figure 3.4—Risk Categories and Questions to Consider

Risk Category	Description	Questions
IT benefit/value enablement risk	Associated with missed opportunities to use technology to improve efficiency or effectiveness of business processes or as an enabler for new business initiatives	• Will the cloud program provide a user environment that matches or exceeds the experience offered by competitors? Will current and new users welcome and embrace the experience? • Will the cloud program be completed and then maintained within estimated budgets, providing or exceeding the expected value?
IT program and project delivery risk	Associated with the contribution of IT to new or improved business solutions, usually in the form of projects and programs as part of investment portfolios	• Can the enterprise's IT organization execute the planned new cloud program successfully, on budget and on schedule? Will the cloud program provide the expected business benefit? Would not moving to the cloud cause additional levels of risk? • Will the cloud program provide a user environment—for internal, partner and customer users—that is well-thought-out, intuitive and welcoming to use? • Will the environment address the prespecified business needs and be able to accommodate requirements identified after implementation?
IT operations and service delivery risk	Associated with all aspects of the business-as-usual performance of IT systems and services, which can bring destruction or reduction of value to the enterprise	• Will the operational support provided by the CSP maintain expected performance and allow for cost-effective scale to support business growth objectives? • Can the new cloud program meet current and future regulatory requirements?

Some enterprises find it useful to consider risk in terms of risk scenarios. A risk scenario is a description of a possible event that, when occurring, will have an uncertain impact—positive or negative—on the achievement of the enterprise's objectives. For example, in considering the selection of cloud technology, a positive risk scenario might be "Optimal technology is selected for implementation" while a negative risk scenario might be "Wrong technologies, in terms of cost, performance, features or compatibility, are selected for implementation." Well-developed risk scenarios support the activities of identifying, analyzing and acting on risk, and help make potential outcomes realistic and relevant to the enterprise.

Risk scenarios can be derived via two different mechanisms:

- **Top-down approach**—This approach, which starts from the overall enterprise objectives and proceeds to an analysis of the most relevant and probable IT risk scenarios impacting the enterprise objectives. If the impact criteria used during risk analysis are well aligned with the real value drivers of the enterprise, relevant risk scenarios will be developed.
- **Bottom-up approach**—This approach, in which a list of generic scenarios is used to define a set of more relevant and customized scenarios, applied to the individual enterprise situation.

These approaches are complementary and should be used simultaneously. Risk scenarios must be relevant and linked to real business risk. Using a set of example generic risk scenarios could assist in identifying risk, reduce the chance of overlooking major/common risk scenarios and can provide a comprehensive reference for IT risk. However, specific risk for each organization and critical business requirements will need to be considered in the organization's risk scenarios.

Simply determining risk, asking pertinent questions and developing risk scenarios are not enough, however, to keep the enterprise at an approved level of risk tolerance. A number of activities must be undertaken to manage risk. A few of those activities, which should constitute part of a fully developed Enterprise Risk Management (ERM) program, include:

- The enterprise must ensure that the results of risk assessments are addressed in risk action plans, which are, in turn, reflected in SLAs.
- A system is established to ensure that risk acceptance is approved by a member of management with the authority to accept the risk on behalf of the organization and who understands the implications of the decision.
- Risk management controls are put in effect in the enterprise and the CSP to manage risk-based decisions.
- The CSP is asked to make available to customers independent third-party assessments to describe the control practices in place at the service provider's operating locations.
- The CSP is asked to submit third-party reviews that certify that the controls have been tested using recognized selection criteria.
- The deployment scenario (SaaS, PaaS, IaaS) defines the data protection responsibilities between the customer and service provider, and these responsibilities are clearly established contractually.

3.5 Cloud Vendor Selection and Management

The selection and ongoing management of a CSP may be the most critical aspects of successful cloud utilization. A CSP that operates in a way that is compatible with the enterprise's needs can ensure a mutually beneficial business relationship. The enterprise plays its part by establishing effective selection and management processes to secure the right vendor, outline performance expected from that vendor, and monitor performance consistently.

3.5.1 Vendor Selection

Once an enterprise has determined that a move to the cloud would be advantageous and selected the most suitable service type and model, it must select a CSP. This is not always an easy task as it entails finding a vendor that best serves the business needs while minimizing potential risk.

A transition to the cloud will be much easier when the CSP has the same industry certifications as the enterprise. This also goes for the business needs of the enterprise: Enterprises that are changing rapidly will want to choose a CSP that is agile and can change quickly as well. If, for example, computing requirements change daily, the enterprise needs a CSP that can accommodate those changes rapidly and fluently. A CSP that can change processor power only biweekly and that also requires the completion of four different forms for each change may not be the best choice for the enterprise.

While there are many suitable CSPs, it is important to choose an established one with the proper references. Established CSPs are potentially more experienced with running cloud infrastructures, adapting to change and generally responding faster to an incident or threat, thus being able to maintain stability more efficiently. However, a larger CSP may also be stricter about its offerings, which could greatly reduce the possibility of fine-tuning services not fully compatible with the enterprise needs.

The following characteristics and challenges can influence the selection of a CSP, depending on the objectives and goals established as part of the governance framework:

- **Vendor expertise**—The enterprise needs expert knowledge or a broader perspective and experience with similar enterprises to effectively and efficiently handle certain activities.
- **Vendor capacity**—If the enterprise does not have the resources to handle the work related to a specific product or service, the vendor can supply the resources to support the entire operation or supplement in-house resources.

- **Vendor assuming risk**—The enterprise wishes to outsource activities to leverage a vendor's experience with operational risk and corresponding risk mitigation services. However, the accountability for the adequate performance of those activities can never be delegated and stays with the enterprise.
- **Vendor leveraging scale**—Vendors can offer services at a lower cost because working for multiple customers allows vendors to leverage scale.
- **Cloud security policy/procedure transparency**—Some CSPs may have less transparency than others when it comes to their current information security policies. The rationalization for this is that the policies may be proprietary. This practice may cause conflict with clients' information compliance requirements. It is important for the enterprise to ascertain whether the CSP has adequate security controls/detection systems in use, such as penetration detection. If such a system is in use, it is important to ensure that it has the required sophistication to monitor all cloud computing activities adequately. It is also important to consider whether a real-time digital dashboard is provided to user managers, along with audit logs and records of security incidents.
- **Legal and regulatory requirements**—For the many compliance requirements, including privacy and PII laws, Payment Card Industry (PCI) requirements and various financial reporting laws, today's cloud computing services can challenge various compliance audit requirements currently in place.
- **Vendor business viability**—As with enterprises in all areas of business, some CSPs will thrive and others will go out of business. Clients need to consider the risk and how data and applications can be easily transferred back to the traditional enterprise or to another CSP.
- **Screening of other cloud computing clients**—By definition, CSPs leverage their cloud computing technology for many clients concurrently to maximize revenue. Clients should consider whether the other clients who share the same servers—and, in the case of SaaS, the same application and data files—are of the same repute as their own enterprises. This degree of due diligence should also apply to vendors that are in the critical path of the CSP. Background checks on these enterprises are important to ensure that data are not being hosted by an organization that is incapable of responding to outages or providing business continuity or that is engaging in malicious or fraudulent activity.
- **Cloud data ownership**—Some CSPs require contracts that provide them ownership of the data placed in the cloud computing environment they maintain. They may also require significant service fees for data to be returned to clients if or when a cloud computing services agreement is terminated.
- **Record protection for forensic audits**—Clients must also consider the availability of data and records, if required for forensic audits. Since data may have been commingled and migrated among multiple servers located widely apart, it may be possible that the data for a specific point in time cannot be located. Furthermore, local authorities may impound a cloud computing server to assess court-warranted data records of a suspect client, taking all the data in the cloud.

Vendor location (storage and processing centers) is of key importance for legal reasons. Laws are applicable locally, which can result in a multitude of different law systems applicable to data or business processes stored in the cloud. The most important applicable laws are:

- **Local law of the enterprise**—Storing data in the cloud does not waive the enterprise from the legal obligations it has toward its customers, employees and shareholders. Therefore, local laws will still apply even if data are stored abroad.
- **Local law of the CSP**—Since the CSP has legal obligations toward its customer (the enterprise), the local laws of the CSP may become applicable to its data centers and its content, including the data or business processes of the enterprise.
- **Local law where data are stored**—Local laws apply to the CSP's storage centers. The country that houses the storage center also imposes the law. This is especially important in relation to privacy. Privacy regulations that the enterprise is bound to may not be applicable in the country where the data are housed, thus potentially compromising the stored data.

- **Local law where data are processed**—In some countries local laws do not merely apply to where data are stored, but also are applicable to the location where data are processed.

More established CSPs will take legal matters into account to avoid legal issues when deploying their data centers and storage centers. It is important, however, to have the CSP's storage regions backed up by an independent organization so it can be assured that data or business processes are in the best possible hands.

Cloud computing requires oversight from the CSP and the client. Enterprises should consider CSPs as partners who share the responsibility of managing assets residing in the cloud. The criticality of these partners will depend on the nature of the assets the enterprise is willing to move to the CSPs environment. The enterprise should remain the sole proprietor of the assets, while the CSP will become a guardian responsible for the activities agreed on in the contract and SLAs.

A CSP is not the only cloud-related vendor the enterprise may wish to research and evaluate. A cloud service broker (CSB) is a third-party provider with access to multiple data centers and cloud service offerings. A CSB integrates and tailors those various services into one service that can be intertwined with enterprise in-house applications and systems. This integration enables the CSB to guarantee interoperability among various cloud services and can make the transition to the cloud more cost-effective.

3.5.2 Vendor Management

An effective vendor management process with goals and objectives can help ensure the following outcomes:

- The vendor selection and management strategy are consistent with enterprise goals.
- Effective cooperation and governance models are in place.
- Service, quality, cost and business goals are clear.
- All parties perform as agreed.
- Vendor risk is assessed and properly addressed.
- Vendor relationships are working effectively, as measured according to service objectives.

These outcomes are critical to the success of the cloud program and involve high-level, coordinated performance of multiple tasks on a consistent basis. Consequently, vendor management is not the work of a single individual or small group. There are responsibilities at virtually every level and function of the enterprise, and they must be performed well to realize the desired outcomes.

The specific responsibilities will vary by enterprise, given different industries, laws, budgets, staffing and organizational structures. However, some general responsibilities can be outlined for various functions, as shown in **figure 3.5**.

Figure 3.5—Responsibilities of Vendor Management Roles

Position	Description	Responsibilities
C Suite	The C-level executive who is accountable for the vendor management process depends on the scale of outsourcing. For involvement of large-scale vendor services covering or impacting the entire enterprise, the chief executive officer (CEO) and chief financial officer (CFO) are closely involved and take on the accountability for the activities described for the CIO. The CEO and CFO should be informed of vendor performance during the operations phase.	CIO: • Often accountable for the execution of the contractual vendor relationship life-cycle phases (when the scope and impact of products and services are limited to IT). • Provides oversight to ensure that contract objectives do not conflict among vendors, products and services; contracts comply with internal policies; contracts are managed properly; and SLAs are met. • Provides oversight and coordination of all responsible parties and supports other C-level executives in the decision-making processes (when the scope also includes the business). CFO: • Responsible for the enterprise budget after it is approved by the board; therefore, should approve the budget available for IT vendors. • Should be at least consulted for drafting the call for tender to validate budget specifications, negotiating with multiple vendors and agreeing on the final vendor contract.
Business process owners	Because the goal of contracting IT products or services is to enable the enterprise to achieve its goals, the business process owners should be actively involved in the vendor management life cycle, in addition to IT and procurement.	• Provide business requirements. • Consolidate stakeholder requirements and make sure they are validated and approved by all parties. • Evaluate potential vendors using specific business knowledge and expertise. • Provide input for contract specifications (e.g., service levels and transition-out specifications).
Procurement	Involving sourcing professionals from the beginning of the process enables them to support business units, IT and compliance during the selection and contract management processes.	• Help business process owners consolidate business requirements from various stakeholders. • Help IT consolidate IT-related requirements. • Prepare and send out the call for tender. • Gather vendor offers. • Facilitate the vendor evaluation process, vendor negotiations and contract management activities.
Legal	To effectively mitigate vendor-related risk, the legal function should be involved throughout the entire vendor management life cycle.	• Provide input regarding service requirements, from a legal point of view. • Consolidate legal requirements and make sure they are validated and approved by the other stakeholders. • Provide boilerplate standard contract language for important legal and compliance provisions. • Evaluate potential vendors using specific legal knowledge and expertise. • Draft the contract, taking into account and reviewing the specifications provided by other stakeholders.

Figure 3.5—Responsibilities of Vendor Management Roles *(cont.)*

Position	Description	Responsibilities
Risk function	The risk function should be consulted throughout the vendor management life cycle to obtain a complete view on risk that is related to the relationship, services or products. After the risk assessment is completed, the enterprise identifies controls required to minimize newly identified risk and updates related documentation (e.g., business impact analyses, BCPs, DRPs, insurance policies and the risk framework).	• During the setup and contract phases, provide requirements to mitigate potential risk. • Address potential risk during the operations phase within the enterprise risk framework, based on information provided by the business units and the enterprise risk appetite. • If the vendor environment, business model, products or services change, initiate an assessment to identify and address new risk.
Compliance and audit	The compliance and audit functions should be consulted throughout the vendor management life cycle to ensure compliance with internal and external laws, regulations and policies.	• Ensure compliance of vendor candidates during the selection process. • Provide compliance requirements to be included in the contract. • Ensure compliance of the selected vendor during operations.
IT	All stakeholders share the responsibility of the vendor management process to ensure that products and services received from the vendor support enterprise goals within risk tolerances. However, the IT role is significant because its members may be more familiar with the products and services and their market availability.	• Identify potential vendors, products and services. • Provide input regarding requirements for products and services from an IT point of view. • Provide input associated with baseline standard requirements and architectural limitations. • Provide expectations associated with good practices and defined architectural boundaries. • Provide information practices, such as information ownership and co-location requirements, partnership arrangements and employee checking. • Consolidate the previous inputs and ensure that they are validated and approved by all stakeholders. • Evaluate potential vendors using specific IT knowledge and expertise. • Provide input for contract specifications (e.g., transition-in specifications, transition-out specifications and service metrics). • Monitor, report and manage vendor performance during operations.
Security	Although security is often associated with the IT stakeholder, security is regarded as a separate stakeholder because it does not always reside within the IT function and can be involved with other functions as well as IT.	• Provide input regarding security requirements for products and services. • Provide input associated with baseline security standard requirements. • Provide information practices, such as information accessibility and required assurances of information protection. • Verify compliance of the vendor with product, service and information security requirements.
Human resources		• Should be consulted throughout the vendor management life cycle to ensure compliance with the enterprise's worker statutes, local regulations, code of conduct and labor law.

However, the client enterprise may be structured, there are generally two main approaches/techniques the client may use to manage and measure a CSP's quality of service performance:

- Use vendor management, including but not limited to vendor risk assessment, vendor due diligence, vendor-tiering (based on the significance of the process outsourced), and contracting and SLAs (as they apply to a CSP).
- Apply independent assurance by either the CSP's auditors or client personnel outside of vendor management (involves understanding the scope of services provided, obtaining and evaluating third-party assurance reports, evaluating residual risk, and determining whether an onsite visit to the service provider is required).

Service Level Agreement

A key element in any vendor management program is the SLA. A well-designed SLA and mature service level management maximize the effectiveness of the services they target and manage. This improves business performance and fosters a better relationship between the enterprise and the vendor through:

- Better alignment with business objectives
- Ability to manage services proactively
- Greater transparency of service delivery
- Lower service level management overhead
- Better relationships between the enterprise and vendor

An SLA is an agreement between the enterprise and a product or service provider. This service provider may be an external party—in the case of this publication, a CSP; an internal department, e.g., the IT department for IT services; or a combination of both. An SLA is very common in IT contracts and can be an effective tool to create a common understanding between contracting parties regarding services, priorities and/or responsibilities in the framework of an agreement.

The purpose of the SLA is to ensure alignment between IT-enabled services and service levels, enterprise needs and expectations regarding the handling, usage, storage and availability of information, and requirements for business continuity and disaster recovery. It facilitates the relationship between the client enterprise and the CSP by making clear the goals and objectives of the relationship, the expected performance levels, the responsibilities of each party, and actions to be taken in the event of nonperformance or termination.

SLAs must be measurable (using process metrics, which inform the CSP and the enterprise of the effectiveness and efficiency of key actions done while delivering service and technology metrics, which provide efficiency data on the component level), to allow meaningful reporting among the parties—reports that can be used during consultations between parties about the achievement of the SLA. Agreed, realistic terms in the SLA can help eliminate grounds for disputes between parties.

The enterprise must determine in advance the terms to be included in the SLA, keeping in mind that strict or complex SLAs may result in higher maintenance cost. Some terms that should be negotiated and documented in the SLA include:

- Availability
- Response time for additional computing resources requests
- Response time for incidents
- Backup policies
- Data retention and disposal policies and procedures
- Path management
- Security controls

- Recovery and continuity objectives
- Controls to satisfy legal and compliance requirements

It is also important to address in the SLA the consequences of failure to meet the performance levels specified in the agreement, perhaps by outlining a set of actions to deal with the cause of such failure. The parties could define penalties in case of nonachievement of the levels established (perhaps coupled with incentives for achievement). If a price correction is defined as a penalty, it is important to stipulate that the price correction is not the only compensation that can be claimed in a case of nonachievement of the SLA, so that the right to the payment of full damage is not jeopardized. For continuity of the enterprise, it is recommended to provide a special clause that states that the enterprise can appoint a third-party service provider to deliver the necessary services in case of nonachievement of the SLA by the service provider.

An audit clause can be included in the contract associated with the SLA to grant the enterprise the right to access relevant financial data, accounting information, operational and technical data, safety standards, etc., of the other party. Use of an audit clause is strongly recommended because it allows the enterprise to check, among other areas, the co-contracting party's performances, compliance and achievements.

Service Level Monitoring

Clearly, the enterprise's focus on service levels does not conclude with the execution of the contract. Consistent, thorough monitoring of service levels must take place for as long as the relationship with the CSP is in place, in accordance with the reporting requirements, performance expectations, penalties for shortfalls in performance and rewards for performance that exceeds expectations, as specified in the SLA. The quality of the monitoring depends, in large part, on open and honest communication between the enterprise and the CSP, as well as a shared commitment to meeting objectives measured via specified metrics.

Monitoring may be done by the client enterprise or it can be turned over to a third party that specializes in that activity. In either case, it is likely that one or more of the multiple monitoring tools on the market will be employed. Given the potential need for a third party and the almost certain need for tools, the enterprise must be aware, before entering the cloud space, that there are costs associated with monitoring CSP activities as they endeavor to deliver on the promises made in the SLA.

Identity and Access Management and Vulnerability Management

Many of the risk concerns associated with cloud management were covered in section 3.4. However, two topics merit additional discussion here because, when it comes to risk, security, incident response, etc., clients are limited by the vendor to what the original contracts allow. These same issues arise with cloud-provided IAM and the ability to conduct or receive vulnerability management reports. The enterprise has limited capabilities to mitigate or respond and must rely on what the vendor is willing to provide. This calls for prudent and effective vendor management.

Identity and Access Management

IAM encompasses people, processes and products to identify and management of the data used in an information system to authenticate users and grant or deny access rights to data and system resources.

Its goal is to provide appropriate access (or deny inappropriate access) to enterprise resources. The enterprise should have a strong interest in ensuring that its CSP has developed and implemented adequate user access privilege controls. For example, the enterprise should ensure that user provisioning (on-boarding), deprovisioning (termination) and job function changes of cloud-based applications and operating platforms are managed in a timely and controlled manner, according to internal user access policies.

The increasingly sophisticated applications offered online and available for access by enterprise users, partners and clients call for the use of highly granular and least-privilege-based user access tools. The responsibilities for ensuring an effective IAM system, as categorized by the CSP and the enterprise, are shown in **figure 3.6.**

Figure 3.6—Responsibilities for IAM

CSP	Enterprise
• Manage user identity and logical access. • Manage physical access to IT assets. • Manage roles, responsibilities, access privileges and levels of authority. • Ensure traceability of information events and accountabilities. • Monitor the infrastructure for security-related events.	• Define requirements. • Communicate requirements. • Choose the level or package of service that best meets enterprise goals (based on budget, risk appetite and anticipated benefits delivery). • Document requirements in contracts and SLAs. • Implement internal processes to manage identity and access requests. • Assess control environment. • Report results.

Source: ISACA, *Controls and Assurance in the Cloud: Using COBIT® 5*, USA, 2014, *https://www.isaca.org/bookstore/cobit-5/cb5ca*

The enterprise's needs and expectations should be spelled out in the contract with the CSP (and managed and monitored consistently thereafter). The agreement should define who can access the enterprise's information, naming specific roles for CSP employees and external partners.

The enterprise should request that the CSP provide detailed technical specifications of its IAM system for the enterprise's chief information security officer (CISO) to review and approve. If the CSP declines to provide such details due to its own internal security policies, the enterprise can request controls and benchmarks, such as the result of penetration testing on the CSP's IAM systems, as an alternative.

An alternative approach is to use the enterprise's IAM systems instead of the CSP's IAM systems. In this way, IAM remains the responsibility of the enterprise, so no access to assets can be granted without the enterprise's knowledge. This approach requires the approval of the CSP and the establishment of a secure channel between the CSP infrastructure and the enterprise's IAM system.

Vulnerability Management

Vulnerability management is another area of special concern to enterprises that use cloud computing services. Information assets could be accessed by unauthorized entities due to faulty or vulnerable access management measures or processes. This could result from a forgery/theft of legitimate credentials or misuse of a common technical practice (e.g., administrator permissions override). The number of new attacks increases regularly and at breakneck speed; it is a constant challenge to stay ahead of the attackers.

Enterprises have a vested interest in ensuring that the CSP is exercising all reasonable efforts to keep the environment in which the enterprise's data are stored and processed free from vulnerabilities that provide opportunities for attackers. Vulnerability management is a key component of a strong information security program, in the enterprise and in the CSP's environment.

However, it is the very nature of the cloud—the fact that organizations do not own and control the infrastructure that provisions the cloud service, nor do they have access to vulnerability data of cloud native services—that make vulnerability management especially difficult. This is coupled with the enterprise's inability to apply security controls or other mitigation to the cloud provider's infrastructure. Ultimately, the enterprise's ability to make informed, risk-based decisions about cloud services is affected.

This is not to say that there is nothing the enterprise can do about potential cloud vulnerabilities. As with IAM, the first step is attention to the contract with the CSP. The enterprise should require the CSP to undertake continuous vulnerability assessment of its infrastructure and provide the enterprise reports that meet the enterprise's specifications for content and frequency.

If the CSP provides the versioning of the software (if there is one), that too may help the enterprise determine whether certain vulnerabilities may exist. The SLA should contain detailed specifications about vulnerability classification and actions taken according to the severity level (and in alignment with the enterprise's policies and procedures for similar events) and clarify who is allowed to access the enterprise's information, naming specific roles for CSP employees and external partners. And, of course, the enterprise should request that the CSP implement application firewalls, antivirus and anti-malware tools. If the request is denied, it may be time to begin a search for a new CSP.

3.6 Portability of Services

Despite the best efforts at vendor selection and management, situations can arise in which it is advisable or necessary to migrate data, applications or platforms from one CSP to another or bring services back in-house. For example:

- The CSP may have applied a price increase the enterprise believes to be disproportionate to a service provided or out of line with marketplace values.
- The CSP may have fallen short of SLA levels of performance or breached enterprise confidentiality.
- The CSP may change its services often, without prior notice to the enterprise.
- The enterprise may have uncovered a new business need that necessitates moving cloud-based resources to a different geographic area, under different jurisdiction.

Portability is a difficult process and can result in business disruption or failure for the client in any case, but especially if the CSP has gone bankrupt, faces legal action or is the potential target for an acquisition (which may entail sudden changes in CSP policies and agreements in place).

This underscores the need to outline clearly in the contract and/or SLA the terms that will trigger the retrieval of the enterprise's assets, the processes surrounding migration, the responsibilities of each party and the time frame for completion of migration as required by the enterprise.

However, even with clear language in the SLA, a smooth, effortless portability process is not guaranteed. Different environments are likely to have compatibility issues, and containers—sometimes touted as a solution to portability woes—cannot always smooth the path. Indeed, some IT experts believe that completely trouble-free portability is an outcome that is unlikely ever to be reached.

For several years, various efforts have been underway to develop standards related to the cloud, including portability. Some activities have been government-led, aimed at creating pertinent legislation. Other efforts have been driven by organizations such as the Institute of Electrical and Electronics Engineers (IEEE) and International Organization for Standardization (ISO)/International Electrotechnical Commission (IEC), which has produced ISO/IEC 19941:2017 *Information technology – Cloud computing – Interoperability and portability*. As standardization becomes more widely accepted and adopted, it should make portability easier, helping cloud computing enable use of the same applications in different environments.

3.7 Chapter 3 Knowledge Check

REVIEW QUESTIONS

1. One reason to implement sound governance with a cloud implementation is to:

   A. Ensure continuity of critical business processes that now extend beyond the data center
   B. Manage increasing risk while ignoring security and other aspects
   C. Prove to management that migration to the cloud is worth it
   D. Move assets off premise to reduce accountability

2. Governance is likely to achieve its mission without complete buy-in from the board of directors.

   A. True
   B. False

3. What is the **BEST** cloud capability to address the business need of cost savings?

   A. Cloud computing makes patch management more flexible.
   B. Cloud computing can provide on-demand implementation in real time.
   C. Cloud computing is self-healing.
   D. Cloud computing can reduce the number of workstations in the office.

4. All of the following are informational asset risk events, except:

   A. Loss
   B. Unavailability
   C. Manipulation
   D. Disclosure

5. Which of the following reflects a governance objective to consider during vendor selection?

   A. Bandwidth
   B. Culture
   C. Processing capability
   D. Risk assumption

6. What provision can be included in an SLA with a vendor to ensure achievement of proper compliance?

   A. Security controls
   B. Backup policies
   C. Retention of data
   D. Audit clause

7. Which of the following **BEST** describes portability of services?

   A. The ability of the CSP to physically move data from one datacenter to another
   B. How the CSP handles billing requirements
   C. The ease of moving data and services from one CSP to another
   D. The number of mobile data centers a CSP provides

Answers on page 54

Chapter 3 ANSWER KEY

Review Questions

1. **A. Ensure continuity of critical business processes that now extend beyond the data center. Refer to page 34.**
   B. Manage increasing risk while ignoring security and other aspects
   C. Prove to management that migration to the cloud is worth it
   D. Move assets off premise to reduce accountability

2. A. True
   B. False. Refer to page 35.

3. A. Cloud computing makes patch management more flexible.
   B. Cloud computing can provide on-demand implementation in real time.
   C. Cloud computing is self-healing.
   D. Cloud computing can reduce the number of workstations in the office. Refer to page 35.

4. A. Loss
   B. Unavailability
   C. Manipulation. Refer to pages 38-39.
   D. Disclosure

5. A. Bandwidth
   B. Culture
   C. Processing capability
   D. Risk assumption. Refer to page 44.

6. A. Security controls
   B. Backup policies
   C. Retention of data
   D. Audit clause. Refer to page 50.

7. A. The ability of the CSP to physically move data from one data center to another
   B. How the CSP handles billing requirements
   C. The ease of moving data and services from one CSP to another. Refer to page 52.
   D. The number of mobile data centers a CSP provides

Chapter 4: Cloud Service Supports

Cloud Service Supports

Cloud Service Supports

4.1 Learning Objectives

After completing this chapter, learners will be able to:

1. Distinguish between service implementation and support in the cloud.
2. Describe the testing and validation requirements for post-cloud implementation.
3. Articulate the special role that configuration management plays in cloud computing.
4. Identify resource management challenges with cloud computing implementation.

4.2 Overview

Section 1.4 discussed cloud implementation from a high-level view and covered topics such as establishing teams, migrating data and conducting post-implementation reviews—all of which could apply to any "typical" in-house IT implementation. In some ways, it is like traditional infrastructure; cloud is, in fact, hardware and software and is, therefore, subject to many, if not most, of the same concerns as the enterprise's in-house systems.

But cloud is not exactly like an in-house implementation. While the technology may be the same, the business model is different. Its very essence—the fact that the enterprise lacks authority over its implementation, controls, testing, business practices, compliance activities, etc., and is dependent on the provider to perform up to the enterprise's expectations—causes it to present unique challenges that require special attention.

4.3 Testing and Validation

Enterprises are accustomed to testing and validating their own systems. They test resilience, security, controls, compatibility and a host of other elements of the technology, the programming and the users who operate it, with the purpose of checking that a system meets specifications and fulfills its intended purpose. In validating their own systems, enterprises freeze the systems, and any intermittent changes requested are implemented under a strict change control process.

This sort of testing is unavailable to enterprises in the cloud. Enterprises cannot freeze and control the changes happening to the system. Changes and improvements are continuous, necessitating a process of continuous validation to be sure that those changes are not negatively impacting the state of the system. Enterprises must be assured that cloud-hosted applications are successfully qualified and in a validated state for the entire duration of their life cycle, and they remain in compliance with the regulating bodies.

The testing and validation process must be negotiated through the SLA and tracked through monitoring. And, while CSPs are generally on board with testing, they may be reluctant to share all aspects of it due to the proprietary nature of some products or the security concerns that may arise from revealing processes.

The enterprise must fully understand the nature of its data being placed in the cloud to ensure proper validation:

- Is it PII?
- Does it have specific compliance requirements, such as those associated with Payment Card Industry Data Security Standard (PCI DSS), the US Health Insurance Portability and Accountability Act (HIPAA), the US Sarbanes-Oxley Act (SOX) or the European Union's General Data Protection Regulation (GDPR)?

The answers to these questions can inform the enterprise's discussions with the CSP and inspire other questions the CSP should address, such as:

- Can a test period be provided for new application candidates to identify possible issues early in the deployment?
- If data are demanded by authorities, can the data be provided without compromising other information?
- What sort of compliance testing is done? How often?
- When change plans are presented, does the CSP perform compatibility testing?
- Is the backend secure and does the enterprise have authorization to run tests on it? If so, how?
- Are business continuity and disaster recovery plans well documented and tested?
- Is segregation of duties between provider employees maintained? How is different customers' information segregated? What controls are in place to prevent, detect and react to breaches?
- How often is backup testing done? Are the procedures for doing the testing consistent with the enterprise's security policy?

Validation processes conducted by the CSP should be supported by a validation plan, current versions of all requirements documents and any qualification testing documents. Validation testing will ensure that all predefined requirements and specifications are successfully challenged and met, and testing is properly documented.

The basic principles of validation, which must be considered when tackling the validation of a cloud-based system, are:[14]

1. Fitness for the intended use of a computerized system is verified and documented.
2. Risk associated with using the system is mitigated.
3. The validated state is maintained through effective change control mechanisms.

Enterprises will be especially interested in the supplier assessment, which determines the provider's ability to meet the enterprise's expectations surrounding controls to ensure security, data integrity, compliance with the enterprise's procedures and regulatory compliance. Audits—both offsite and onsite—can provide this information.

Ultimately, the validation of cloud-based systems should align with the risk level of the applications.

4.4 Configuration Accessibility

Configuration is an important element of cloud computing because mistakes made in configurations can offer opportunities for attackers to access data. The types of misconfigurations that can result in attacks can be categorized as follows:[15]

- **Storage access**—Enterprises may have false confidence in the security surrounding their cloud storage because they believe access is afforded to "authenticated users" only, a term they believe applies only to those who are authenticated with their organization or the relevant application. However, when speaking of Amazon Web Services (AWS), the term "authenticated users" refers to anyone with an AWS authentication: effectively, any AWS customer. Controls must be established to ensure that storage access is limited to only those within the enterprise who need access.
- **"Secrets" management**—This refers to secrets such as passwords, API keys, administrator credentials and encryption keys. Misconfiguring secure storage of this information can reveal it to those who would use it with negative intent. One solution is to maintain an inventory of all secrets the enterprise uses in the cloud and checking it regularly to see how each is secured. A secrets management system can be helpful as well.

[14] Mitchell, T.; "Why You Need to Validate (and Re-Validate) a Cloud-Based System," Montrium, 8 January 2020, *blog.montrium.com/experts/why-you-need-to-validate-and-re-validate-a-cloud-based-system*

[15] Smith, P.; "5 Common Cloud Configuration Mistakes," DARKReading, 17 September 2019, *www.darkreading.com/cloud/5-common-cloud-configuration-mistakes/a/d-id/1335768*

- **Disabled logging and monitoring**—Public clouds provide logs and telemetry data that can be useful to an enterprise, but many enterprises fail to enable, configure or use them. Enterprises should assign responsibility to someone on the cloud team to regularly review the information and flag security-related events.
- **Overly permissive access to hosts, containers and virtual machines**—Enterprises must secure important ports and disable—or at least lock down—older, insecure protocols in the cloud, just as they would in their on-premises data center.
- **Lack of validation**—Whether an internal resource or an outside auditor is used, enterprises must assign someone responsibility for regularly (on a set schedule) verifying that services and permissions are properly configured and applied. Cloud configurations should also be audited periodically.

In addition to these potential misconfigurations, enterprises must give attention to configuration accessibility. Many cloud solutions offer multiple services but not every enterprise needs or wishes to use them all. For some enterprises, taking the entire configuration as offered may be the right decision; others may decide they can address certain needs (e.g., a firewall function) on their own or with a third party.

Effective configuration management requires that the management and provisioning procedures are segregated from the CSP, are limited to a security operations function within the enterprise and provide audit trails to document all activities.

4.5 Management of Resources

Cloud services are rented, not owned, and are paid for on a sliding scale. The compensation corresponds to the number/amount of resources (CPU, memory, energy consumption, etc.) used. Therefore, it is in the enterprise's best interest to take a keen interest in managing the use of those resources at every phase of their life cycle (discovery, allocation, scheduling and monitoring). Inefficient resource management has a direct negative effect on performance and cost, while also indirectly affecting system functionality.

Resource management, which is a core function in any cloud system, is not entirely about cost containment, however. Because of the nature and structure of the cloud, there are other concerns related to resource management, such as identity management, security, availability and privacy.

The policies for resource management can be grouped into a few general categories:[16]

- **Admission control**—To prevent the system from accepting workload that violates high-level system policies and prevent the system from completing work already in progress or finalized
- **Capacity allocation**—To allocate resources for individual instances, a feat that is more difficult when the state of individual servers changes rapidly
- **Load balancing and energy optimization**—To keep the workload at an efficient level to maintain a certain cost of services
- **Quality of service guarantees**—To address availability and performance of applications and systems and the appropriateness of the applications for accomplishing business tasks

The strategies for resource management will differ depending on the cloud deployment and delivery models used by the enterprise. However, the most common resource management issue is fluctuation of workload. Sometimes a spike can be predicted, as in seasonal activity, in which case the appropriate resources can be provisioned in advance. In other cases, when the spike is unplanned and unforeseen, auto-scaling may be employed, provided there is an available pool of resources that can be released or allocated and a monitoring system is in place to allow a control loop to decide in real time to reallocate resources.[17]

[16] ScienceDirect, "Cloud Resource Management and Scheduling," *www.sciencedirect.com/topics/computer-science/cloud-resource-management*
[17] *Ibid.*

Whatever load-balancing techniques are selected, they must consider the factors/targets summarized in **figure 4.1**.

Figure 4.1—Load-Balancing Factors/Targets

Factor/Metric	Definition	Target
Overhead	• Determines the amount of overhead involved while implementing load-balancing algorithms • Composed of overhead due to movement of tasks, interprocess and interprocessor communication	• Should be minimized so that load-balancing techniques can work efficiently
Throughput	• Used to calculate the number of tasks whose execution has been completed	• Should be high to improve the performance of the system
Performance	• Used to check the efficiency of the system	• Must aim for improvement at a reasonable cost; for example, reducing response time while keeping some acceptable delays
Resource utilization	• Used to check the utilization of resources in a system	• Should be optimized for efficient load balancing
Scalability	• The ability of an algorithm to perform load balancing for a system with any finite number of nodes	• Must aim for improvement
Response time	• The amount of time taken to respond by a particular load-balancing algorithm in a distributed system	• Should be minimal, to improve efficiency
Fault tolerance	• The ability of an algorithm to perform uniform load balancing despite arbitrary node or link failure	• All systems expected to be highly fault tolerant

Source: Mahendran, D.; M. Gopi; S. Priyadharshini; R. Karthick; "A Study on Cloud Management Techniques," *International Journal on Innovative Research in Computer and Communication Engineering*, volume 1, issue 1, 2013, p. 60-61

Cloud computing initiatives tend to impact a wider variety of people inside (and sometimes even outside) the enterprise, compared to other services being outsourced. Most enterprises outsource services that are not core to them, thus impacting a smaller amount of people. However, with cloud computing, enterprises most frequently target their core services and take them to the cloud to enjoy the benefits of cloud computing for their core businesses. Obviously, this tends to impact a significant number of people throughout the entire enterprise, thus requiring adequate preparation and management of all resources to keep ensuring operability.

4.6 Storage and Data Availability

Availability—of storage and data—is a significant driver for enterprises to move certain activities to the cloud. They want a place to process and store their data without having to invest in the infrastructure to do so in-house. Therefore, in selecting a CSP and managing the relationship with that CSP, agreement on availability expectations is critical.

Availability—the amount of time the provider guarantees that the enterprise's data and services are available to the enterprise—is generally documented as a percentage of time per year. For example, a 99.9 percent guarantee from the CSP tells the enterprise that its data/storage will be available to the enterprise from the data center on nearly a full-time, 24/7 basis, or conversely that the enterprise will be unable to access the data for no more than about 10 minutes per week, or less than nine hours per year.

A 100 percent guarantee is not given, as the CSP must plan on downtime for maintenance. This should be acceptable to the enterprise, if the downtimes planned by the CSP do not coincide with the enterprise's need for its data.

(Availability does not include guarantees related to connectivity issues that are out of the control of the provider, such as Internet outages or power losses.)

Each enterprise must determine what level of availability is required for its operation. Enterprises that depend on real-time transactions (such as an e-commerce site) will need a higher degree of availability than enterprises that are simply storing backups. Enterprises processing real-time transactions can attach a very real financial impact to periods of unavailability.

As devastating as the direct financial impact may be, however, other losses that may arise due to unavailability, such as a loss of reputation and customer trust, can be even more harmful, long-lasting and less conducive to amelioration.

The enterprise's CSP should publish and guarantee availability and that guarantee should specify the compensation that will be provided to the enterprise if the provider falls short of the guaranteed availability metrics. The enterprise must be sure to read the SLA closely, to see how the CSP defines "availability." Does the CSP consider it not a downtime if even one Internet client can access even one service? Or does the CSP consider availability to exist only when multiple Internet service providers and countries can access all services? The difference between these two interpretations can be staggering and must be evaluated in terms of the enterprise's needs in achieving business objectives.

4.7 Business Continuity and Disaster Recovery

Most enterprises have detailed plans to deal with disasters or disruptions within "the walls" of the enterprise itself. They should have equally comprehensive plans for their cloud computing service. Indeed, many enterprises consider cloud computing services as backup and a plan for disaster recovery. And, while it can serve that purpose, it requires considerable communication between the client enterprise and the CSP to ensure clear, agreed understanding about the assignment of responsibilities and the expectations for performance.

In preparing BCPs and DRPs, the enterprise should plan for the worst-case scenario: complete disruption of the CSP. Maintaining availability of data should be a core element in the plans and expectations should be clearly spelled out, with recovery point objectives (RPOs) and recovery time objectives (RTOs) defining the data that must be restored and when that data must be accessible for the enterprise to resume operations. (The CSP's RPOs, RTOs, backup testing frequency and procedures must be consistent with the enterprise's security policy.)

Other factors the enterprise should consider in preparing BCPs/DRPs include:[18]

- Maintaining backup copies of enterprise data so failure of a storage component or the deterioration of stored data does not result in permanent loss of the information. This should be supplemented with automatic failover, which will automatically replace an affected component with a backup.
- Creating multiple copies of enterprise data and multiple access routes to the data to avoid inaccessibility if any one network component, storage device or server fails.
- Avoiding the temptation to try to increase data availability through in-house, ad hoc "solutions" rather than simply using tools specifically designed for the purpose.

Of course, the negotiation and agreement phases of initiating a relationship with a CSP should include confirming that the CSP already has disaster recovery and backup procedures in place. Various provisions can be included in the contract to ensure both parties agree about processes and responsibilities in the event of a disaster or disruption, including the following activities:

- Document the technical mechanisms and procedures that prevent data loss, such as contractor/enterprise responsibilities and routines for backup, failover or redundancy.

[18] Precisely, "Data Availability 101: What Data Availability Means and How to Achieve It," 16 June 2020, *blog.syncsort.com/2018/06/data-availability/data-availability-101/*

- Provide for continuity of accessibility, usability and preservation of all agency data, regardless of any migration of data to other formats during the contract. Terms should provide for appropriate testing to ensure data integrity prior to any migration.
- Specify provisions and procedures for backup, restoration of services and disaster recovery.
- Upon transfer of data, ensure that technological parity with other service providers is guaranteed.
- Guarantee the preservation of data and provide for routine monitoring of data to identify formats that are at risk of obsolescence.
- Include provisions relating to the migration of data to new formats, when appropriate, and the provision of proper documentation about migration activities to the enterprise.
- Include provisions for the safe return/transfer of data if the CSP is the subject of a takeover.
- Specify what will happen to the data when the contract terminates; for example, transfer to a new provider, return to the enterprise or permanently delete.
- Specify remedies for service-provider mistakes or breaches.
- Identify penalty provisions imposed by the service provider, such as suspension of enterprise access to the data because of nonpayment.
- Precisely specify the terms for the disposal of specific data during the term, at the request of the enterprise and at the end of the term, including a warranty in relation to technological parity/obsolescence.
- Limit suspension and termination rights that are available to the CSP.
- Allow subscription levels to be scalable up and down according to demand.
- Specify reporting and audit rights of the enterprise and vendor.

While there may be overlap between a DRP and a BCP, the BCP focuses on the responsibilities, processes and activities that ensure sustained execution of the enterprise's business activities within its environment, its infrastructure operations and its IT controls. Most BCPs cover responsibilities, accountabilities and authorities; the frequency and processes involved in testing the BCP; and the overall process itself: identifying and declaring a disaster/interruption, carrying out appropriate internal and external communication, restoring activities, analyzing the disaster/interruption, and declaring closure.

The CSP's role in business continuity is to train its employees to imagine all reasonable disaster and business interruption eventualities and devise alternate methods to restore the cloud services necessary to maintaining customers' business and infrastructure needs, thereby minimizing business impact. In other words, when problems occur at the CSP, the customer should be harmed as little as possible.

Page intentionally left blank

4.8 Chapter 4 Knowledge Check

REVIEW QUESTIONS

1. CSPs usually allow enterprises full access to assets for testing.

   A. True
   B. False

2. The **BEST** way for an enterprise to ensure that it has some access for testing and validation of cloud assets is through:

   A. OLAs
   B. SLAs
   C. Safe-haven agreements
   D. SOC2 reports

3. A cloud deployment for a company that deals with health information of US citizens will need to comply with:

   A. CCPA
   B. GDPR
   C. HIPAA
   D. PCI-DSS

4. All of the following reflect misconfigurations that can lead to attacks except:

   A. Lack of validation
   B. Improper storage access
   C. Lack of audit clause
   D. Disabled logging

5. Cloud servers and services are _____ to the customer by the CSP.

   A. Sold
   B. Rented
   C. Described in detail
   D. Escrowed

6. A target of the performance metric is:

   A. System fault tolerance
   B. Reducing response time
   C. Load balancing
   D. Migration time minimization

7. One of the roles that the CSP plays in business continuity is to:

   A. Ensure customers have well-conceived BCPs and are prepared
   B. Provide services free of charge during disasters
   C. Write BCP clauses into SLAs
   D. Ensure their employees have prepared for eventualities

Answers on page 64

Chapter 4 ANSWER KEY

Review Questions

1. **A. True. Refer to page 56.**
   B. False

2. A. OLAs
   B. SLAs. Refer to page 56.
   C. Safe-haven agreements
   D. SOC2 reports

3. A. CCPA
   B. GDPR
   C. HIPAA. Refer to page 56.
   D. PCI-DSS

4. A. Lack of validation
   B. Improper storage access
   C. Lack of audit clause. Refer to pages 57-58.
   D. Disabled logging

5. A. Sold
   B. Rented. Refer to page 58.
   C. Described in detail
   D. Escrowed

6. A. System fault tolerance
   B. Reducing response time. Refer to page 59.
   C. Load balancing
   D. Migration time minimization

7. A. Ensure customers have well-conceived BCPs and are prepared
   B. Provide services free of charge during disasters
   C. Write BCP clauses into SLAs
   D. Ensure their employees have prepared for eventualities. Refer to page 61.

Glossary

A

Application development sandbox—The use of a standalone computer, virtual machine or virtual environment to conduct software development removed from production infrastructure

Application programming interface (API)—A set of routines, protocols and tools referred to as building blocks used in business application software development

Audit trail—Data in the form of a logical path linking a sequence of events, used to trace the transactions that have affected the contents of a record

Source : ISO

Availability—Ensuring timely and reliable access to and use of information

B

Business continuity plan (BCP)—A plan used by an enterprise to respond to disruption of critical business processes; depends on the contingency plan for restoration of critical systems

Business process owner—The individual responsible for identifying process requirements, approving process design and managing process performance.

Scope Notes**:** Must be at an appropriately high level in the enterprise and have authority to commit resources to process-specific risk management activities

Business-to-business—Transactions in which the acquirer is an enterprise or an individual operating in the ambits of his/her professional activity. In this case, laws and regulations related to consumer protection are not applicable.

Scope Notes: The contract's general terms should be communicated to the other party and specifically approved. Some companies require the other party to fill out check-boxes where there is a description such as "I specifically approve the clauses" This is not convincing; the best solution is the adoption of a digital signature scheme, which allows the approval of clauses and terms with the non-repudiation condition.

C

Capital expenditure/expense (CAPEX)—An expenditure that is recorded as an asset because it is expected to benefit more than the current period. The asset is then depreciated or amortized over the expected useful life of the asset.

Change control— The processes, authorities and procedures to be used for all changes that are made to the computerized system and/or the system data. Change control is a vital subset of the quality assurance (QA) program in an enterprise and should be clearly described in the enterprise standard operating procedures (SOPs).

See Configuration control.

Change management—A holistic and proactive approach to managing the transition from a current to a desired organizational state, focusing specifically on the critical human or "soft" elements of change.

Scope Notes: Includes activities such as culture change (values, beliefs and attitudes), development of reward systems (measures and appropriate incentives), organizational design, stakeholder management, human resources (HR) policies and procedures, executive coaching, change leadership training, team building and communication planning and execution.

Chargeback—The redistribution of expenditures to the units within a company that gave rise to them.

Scope Notes: Chargeback is important because without such a policy, misleading views may be given as to the real profitability of a product or service because certain key expenditures will be ignored or calculated according to an arbitrary formula.

Chief executive officer (CEO)—The highest ranking individual in an enterprise

Chief financial officer (CFO)—The individual primarily responsible for managing the financial risk of an enterprise

Chief information officer (CIO)—The most senior official of the enterprise who is accountable for IT advocacy, aligning IT and business strategies, and planning, resourcing and managing the delivery of IT services, information and the deployment of associated human resources.

Scope Notes**:** In some cases, the CIO role has been expanded to become the chief knowledge officer (CKO)

who deals in knowledge, not just information. Also see chief technology officer (CTO).

Chief information security officer (CISO)—The person in charge of information security within the enterprise

Cloud access security brokers (CASBs)—Software or appliances that are positioned between an enterprise technology infrastructure and a cloud service provider (CSP)

Cloud computing—Convenient, scalable on-demand network access to a shared pool of resources that can be provisioned rapidly and released with minimal management effort or service provider interaction

Community cloud—A cloud computing environment in which resources are shared among entities within shared industries or interests in common, e.g., healthcare or financial services

Compliance—Adherence to, and the ability to demonstrate adherence to, mandated requirements defined by laws and regulations, as well as voluntary requirements resulting from contractual obligations and internal policies

Confidentiality—Preserving authorized restrictions on access and disclosure, including means for protecting privacy and proprietary information

Containers—A packaged environment that includes all necessary dependencies, executables, and code for particular applications to run separate from the host computing device

D

Data life cycle—The sequence of steps data go through, beginning with its collection/generation and ending with archiving or deleting it at the end of its useful life

Data loss prevention—Detecting and addressing data breaches, exfiltration or unwanted destruction of data

Data processing—Any operation or set of operations which is performed on personal data or on sets of personal data, whether or not by automated means, such as collection, recording, organization, structuring, storage, adaptation or alteration, retrieval, consultation, use, disclosure by transmission, dissemination or otherwise making available, alignment or combination, restriction, erasure or destruction.

Disaster recovery plan (DRP)—A set of human, physical, technical and procedural resources to recover, within a defined time and cost, an activity interrupted by an emergency or disaster

Due diligence—The performance of those actions that are generally regarded as prudent, responsible and necessary to conduct a thorough and objective investigation, review and/or analysis

E

Enterprise risk management (ERM)—The discipline by which an enterprise in any industry assesses, controls, exploits, finances and monitors risk from all sources for the purpose of increasing the enterprise's short- and long-term value to its stakeholders.

F

Fault tolerance—A system's level of resilience to seamlessly react to hardware and/or software failure.

G

Governance—The method by which an enterprise ensures that stakeholder needs, conditions and options are evaluated to determine balanced, agreed-on enterprise objectives are achieved. It involves setting direction through prioritization and decision making, and monitoring performance and compliance against agreed-on direction and objectives.

H

Hybrid cloud—A cloud computing environment that combines services and resources from both private and public clouds

Hypercall—A stopgap between a hypervisor and the host to filter and control privileged operations

Hypervisors—Software that allows multiple virtual machines to be run on a host machine or group of hosts machines

I

Identity and access management (IAM)—Encapsulates people, processes and products to identify and manage the data used in an information system to authenticate users and grant or deny access rights to data and system resources. The goal of IAM is to provide appropriate access to enterprise resources.

Information security—Ensures that, within the enterprise, information is protected against disclosure to unauthorized users (confidentiality), improper modification (integrity) and nonaccess when required (availability). Information security deals with all formats of information—paper documents, digital assets, intellectual property in people's minds, and verbal and visual communications.

Infrastructure as a Service (IaaS)—Offers the capability to provision processing, storage, networks and other fundamental computing resources, enabling the customer to deploy and run arbitrary software, which can include operating systems (OSs) and applications

Institute of Electrical and Electronic Engineers (IEEE)—An organization involved in the generation and promulgation of standards. IEEE standards represent the formalization of current norms of professional practice through the process of obtaining the consensus of concerned, practicing professionals in the given field

L

Life cycle—A series of stages that characterize the course of existence of an organizational investment (e.g., product, project, program)

M

Metering—The monitoring and tracking of resource usage within a cloud environment, e.g., data, memory and storage

O

On-demand self-service—The ability for a customer to self-assign and allocate cloud resources instantaneously without vendor interaction

Organizational structure—A component of a governance system. Includes the enterprise and its structures, hierarchies and dependencies

P

Personally identifiable information (PII)—Any information that can be used to establish a link between the information and the natural person to whom such information relates, or that is or might be directly or indirectly linked to a natural person

Platform as a Service (PaaS)—Offers the capability to deploy onto the cloud infrastructure customer-created or -acquired applications that are created using programming languages and tools supported by the provider

Private cloud—An on- or off-premises cloud environment in which a specific enterprise controls all infrastructure resources

Provisioning—Allocating resources for cloud computing infrastructure or instance

Public cloud—A cloud environment in which resources are shared between enterprises and individuals

R

Rapid elasticity—The ability to quickly increase or reduce the amount of resources utilized by a cloud computing instance or infrastructure

Recovery point objective (RPO)—Determined based on the acceptable data loss in case of a disruption of operations. It indicates the earliest point in time that is acceptable to recover the data. The RPO effectively quantifies the permissible amount of data loss in case of interruption.

Recovery time objective (RTO)—The amount of time allowed for the recovery of a business function or resource after a disaster occurs

Resilience—The ability of a system or network to resist failure or to recover quickly from any disruption, usually with minimal recognizable effect

Resource management—1. The coordinated activities to direct and control an enterprise with regard to Scope Notes: In the International Standard, the term "control" is used as a synonym for "measure." (ISO/IEC Guide 73:2002)

2. One of the governance objectives. Entails recognizing risk; assessing the impact and likelihood of that risk; and developing strategies, such as avoiding the risk, reducing the negative effect of the risk and/or transferring the risk, to manage it within the context of the enterprise's risk appetite.

Resource pooling—In cloud computing, the ability to combine computing resources and services to serve multiple customers at once

Risk appetite—The amount of risk, on a broad level, that an entity is willing to accept in pursuit of its mission.

Risk management—1. The coordinated activities to direct and control an enterprise with regard to risk

Scope Notes: In the International Standard, the term "control" is used as a synonym for "measure." (ISO/IEC Guide 73:2002)

2. One of the governance objectives. Entails recognizing risk; assessing the impact and likelihood of that risk; and developing strategies, such as avoiding the risk, reducing the negative effect of the risk and/or transferring the risk, to manage it within the context of the enterprise's risk appetite.

Scope Notes: COBIT 5 perspective

Risk scenario—The tangible and assessable representation of risk.

Scope Notes: One of the key information items needed to identify, analyze and respond to risk (COBIT 2019 objective APO12)

Risk tolerance—The acceptable level of variation that management is willing to allow for any particular risk as the enterprise pursues its objectives

Risk transfer—The process of assigning risk to another enterprise, usually through the purchase of an insurance policy or by outsourcing the service

Scope Notes: Also known as risk sharing

S

Security as a Service (SecaaS)—The next generation of managed security services dedicated to the delivery, over the Internet, of specialized information-security services.

Service level agreement (SLA)—An agreement, preferably documented, between a service provider and the customer(s)/user(s) that defines minimum performance targets for a service and how they will be measured

Service-oriented architecture (SOA)—A cloud-based library of proven, functional software applets that are able to be connected together to become a useful online application.

Shadow IT—The use of systems, services, hardware or software on an enterprise network or within an enterprise's infrastructure without proper vetting and approval from the IT or cybersecurity department

Single sign-on (SSO)—A single point authentication system used by multiple systems and applications

Software as a service (SaaS)—Offers the capability to use the provider's applications running on cloud infrastructure. The applications are accessible from various client devices through a thin client interface, such as a web browser (e.g., web-based email).

Solution stack—A collection of hardware, software and services that work in unison to provide an enterprise or user with a final product

V

Validation—Establishing documented evidence that provides a high degree of assurance that a specific process will consistently produce a product meeting its predetermined specifications and quality attributes

Virtual machine—An emulation of a computing environment or operating system separate from the host computing system

Virtual machine (VM) jumping—Exploitation of a hypervisor that allows an attacker to gain access to one virtual machine from another

Virtualization—The process of adding a guest application and data onto a virtual server, recognizing that the guest application will ultimately be removed from this physical server

9781604208382